Wireless Power Transfer – Recent Development, Applications and New Perspectives

Edited by Mohamed Zellagui

Published in London, United Kingdom

IntechOpen

Supporting open minds since 2005

Wireless Power Transfer - Recent Development, Applications and New Perspectives
http://dx.doi.org/10.5772/intechopen.92531
Edited by Mohamed Zellagui

Contributors
Nassim Iqteit, Khalid O. Moh. Yahya, Sajjad Ahmad Khan, Himanshu Dehra, Rajarshi Mahapatra, Poonam Lathiya, Jing Wang, Sarawuth Chaimool, Prayoot Akkaraekthalin, Yan Zhao, Chawalit Rakluea, Michal Frivaldsky, Vladimir Kindl, Jakub Skorvaga, Mohamed Aboualalaa, Hala Elsadek, Ramesh Pokharel, Mohamed Zellagui, Martin Zavrel

First published in London, United Kingdom, 2021 by IntechOpen
IntechOpen is the global imprint of INTECHOPEN LIMITED, registered in England and Wales, registration number: 11086078, 5 Princes Gate Court, London, SW7 2QJ, United Kingdom
Printed in Croatia

British Library Cataloguing-in-Publication Data
A catalogue record for this book is available from the British Library

Additional hard and PDF copies can be obtained from orders@intechopen.com

Wireless Power Transfer - Recent Development, Applications and New Perspectives
Edited by Mohamed Zellagui
p. cm.
Print ISBN 978-1-83968-801-0
Online ISBN 978-1-83968-802-7
eBook (PDF) ISBN 978-1-83968-803-4

For EU product safety concerns:
IN TECH d.o.o., Prolaz Marije Krucifikse Kozulić 3, 51000 Rijeka, Croatia, info@intechopen.com or visit our website at intechopen.com.

Meet the editor

Dr. Mohamed Zellagui was born in Constantine, Algeria, in 1984. He received a BEng (Hons.) and MSc in Electrical Power Engineering from the University of Constantine 1, Algeria, in 2007 and 2010, respectively. He obtained a Ph.D. in Power Systems from the University of Batna 2, Algeria, in 2014. He was a visiting researcher with various universities in France, Malaysia, Italy, and Canada. He collaborates with many universities and laboratories. He has authored or co-authored more than fifty refereed journal articles and conference papers, five book chapters, and three edited books. He is a member of Conseil International des Grands Réseaux Electriques (CIGRE) and senior member of Institute of Electrical and Electronics Engineers (IEEE). He is also an editorial board member and reviewer for many international journals and conferences. His research interests include power systems, renewable energy integration, energy efficiency, optimization algorithm, wireless power transfer, and smart grids.

Contents

Preface

Wireless Power Transfer – Recent Development, Applications and New Perspectives reviews wireless power transfer (WPT) techniques with an emphasis on fundamentals, technical challenges, metamaterials, and typical applications. It describes WPT technology using magnetic resonant coupling and electric resonant coupling and presents the latest approaches to its practical implementation, operation, and applications.

Power transfer depends on coil coupling, which depends on the distance between coils, alignment, coil dimensions, coil materials, number of turns, magnetic shielding, impedance matching, frequency, and duty cycle. Receiver and transmitter coils must be aligned for best coupling and efficient power transfer.

The book discusses the difference between electromagnetic induction and magnetic resonant coupling, the characteristics of various types of resonant circuit topologies, and the unique features of magnetic resonant coupling methods. It summarizes the key technical issues of WPT systems in terms of efficiency, power, distance, misalignment, directional charging, and energy security. The book also offers a survey on the studies of metamaterial-based WPT systems. It presents the very latest in theory and technology of both coupling and radiative wireless power transfer. Lastly, the book presents some typical applications of WPT technology, including EVs, biomedical implants, and portable electronics.

By reviewing the development and the current state of WPT methods over the last several years, this book offers readers a wide view of WPT techniques based on the inductive coupling effect of the non-radiative electromagnetic field. This emerging energy transmission mechanism has significant impacts on the pervasive application of renewable energies in our daily life.

Designed to be self-contained, this richly illustrated book is a valuable resource for a broad readership, from researchers to engineers and anyone interested in cutting-edge technologies in WPT.

Mohamed Zellagui
Department of Electrical Engineering,
University of Batna 2,
Batna, Algeria

Department of Electrical Engineering,
University of Quebec,
Quebec, Canada

Section 1

Introduction

Chapter 1

Introductory Chapter: Overview of Wireless Power Transfer Technologies

Mohamed Zellagui

1. Introduction

The idea of Wireless Power Transfer (WPT) is not a new concept. From the end of the 19th Century, Nikola Tesla was already experimenting with the possibilities. In recent years, the concept of WPT has gained increased popularity and has raised much excitement [1–3].

Following significant advancements in wireless data transfer over the past two decades, power is the last physical connection to be eliminated. However, the obsession with wireless power is not new.

2. History

The idea dates to 1888 in a work by Heinrich Hertz. He was able to demonstrate high frequency power transfer using a spark gap and parabolic reflectors at both the transmitting and receiving ends of the system.

The more famous wireless power transfer experiment was carried out later by Nikola Tesla in 1893 [1, 2]. He spent thousands of dollars and performed public demonstrations of wireless power transfer, which earned him the title of a mad scientist.

3. Motivation

The technological breakthroughs made possible due to WPT have been fostering an increased research activity, as well as the commercialization of numerous consumer products based on these concepts.

From a research and development perspective, there are two principal fields of investigation in WPT.

The first approach comprises the underlying mechanisms of power transfer, particularly by seeking knowledge about the possible scientific methods to wirelessly transmit energy from a point to another in space.

The second approach includes the development of individual circuit modules required in a functional wireless power transmission system. This includes circuit design considering the specifics of a particular WPT mechanism, ensuring that the module has the appropriate characteristics for that power delivery mechanism.

IntechOpen

4. System overview

There are many different types of WPT such as electromagnetic, electrostatic, electromechanical, etc. In the case of electromagnetic wireless power transfer, the systems are generally broken down into two categories of far-field and near-field power transfer [4].

In such systems, the energy transfer is accomplished through electromagnetic radiation, which for omnidirectional propagation scenario results in poor efficiency. The range that is the far-field region of a transmitter depends on the frequency of operation.

The idea behind WPT is simple. According to the Faraday's law of induction when a varying magnetic field passes through a loop it will produce a potential difference across it. In other words, a loop can harvest an alternating magnetic field which then can be converted and stored as electrostatic energy [5].

Generating, the required magnetic field can be done through Amperes law. Amperes law states that when an electric current flow through a loop, let us call it the transmitter loop, it generates a magnetic field around it.

Now, if we place a second loop, let us call it the receiver loop, in the proximity of the transmitter, an alternating voltage will appear across the receiver loop. Hence effectively transmitting power through electromagnetic waves.

5. Methods

Radio Frequency (RF): RF-based WPT is established as a technology for the transfer of power over the order of tens of meters, however, the efficiency of such systems is much lower than other WPT methods.

Inductive Power Transfer (IPT): This method utilizes the concept that an electric current driven through a coil, by virtue of the produced magnetic flux, induces an electromotive force in an adjacent coil.

Magnetic Resonant Coupling (MRC): Magnetic resonant coupling is deemed as the potential breakthrough in WPT methods [6]. The underlying concept is that two resonator circuits tuned at the same resonance frequency, can experience power transfer at higher efficiencies, at greater distances, when compared with conventional IPT systems.

6. Conclusion

Recent advancements in the semiconductor integrated circuits and functional materials technologies have accelerated the demand for electronic devices such as the internet of things (IoT) and wearable sensors, which have low power consumption, miniature size, and high data transfer efficiency.

Wireless power transfer has become the alternative solution to current electronic devices that rely on bulky batteries to supply the power and energy.

Author details

Mohamed Zellagui[1,2]

1 Department of Electrical Engineering, University of Batna 2, Batna, Algeria

2 Department of Electrical Engineering, University of Quebec, Quebec, Canada

*Address all correspondence to: m.zellagui@univ-batna2.dz; m.zellagui@ieee.org

References

[1] Z. Zhang, H. Pang, A. Georgiadis, and C. Cecati, "*Wireless Power Transfer - An Overview*", IEEE Transactions on Industrial Electronics, vol. 66, no. 2, pp. 1044-1058, 2019.

[2] P. Pinho, "*Recent Wireless Power Transfer Technologies*", Published by IntechOpen, London, United Kingdom, 2020.

[3] A. Triviño-Cabrera, J. M. González-González, and J. A. Aguado, "*Wireless Power Transfer for Electric Vehicles: Foundations and Design Approach*", Power Systems book series, Published by Springer, Cham, 2020.

[4] K. Jin, and W. Zhou, "*Wireless Laser Power Transmission: A Review of Recent Progress*", IEEE Transactions on Power Electronics, vol. 34, no. 4, pp. 3842-3859, 2019.

[5] K. A. Kalwar, M. Aamir, and S. Mekhilef, "*Inductively Coupled Power Transfer (ICPT) for Electric Vehicle Charging – A Review*", Renewable and Sustainable Energy Reviews, vol. 47, pp. 462-475, 2015.

[6] Z. Zhang, K. T. Chau, C. Qiu, and C. Liu, "*Energy Encryption for Wireless Power Transfer*", IEEE Transactions on Power Electronics, vol. 30, no. 9, pp. 5237-5246, 2015.

Section 2

Wireless Power Transfer in Electrical Systems

Chapter 2

Developments in Wireless Power Transfer Using Solar Energy

Himanshu Dehra

Abstract

This chapter presents state-of-the-art and major developments in wireless power transfer using solar energy. The brief state-of-the-art is presented for solar photovoltaic technologies which can be combined with wireless power transfer (WPT) to interact with the ambient solar energy. The main purpose of the solar photovoltaic system is to distribute the collected electrical energy in various small-scale power applications wirelessly. These recent developments give technology based on how to transmit electrical power without any wires, with a small-scale by using solar energy. The power can also be transferred wirelessly through an inductive coupling as an antenna. With this wireless electricity we can charge and make wireless electricity as an input source to electronic equipment such as cellphone, MP3 Player etc. In harvesting energy, technologies of ambient solar radiation like solar photovoltaic, kinetic, thermal or electro-magnetic (EM) energy can be used to recharge the batteries. Radio frequency (RF) harvesting technologies are also popular as they are enormously available in the atmosphere. The energy converted to useful DC energy which can be used to charge electrical devices which need low power consumption. The chapter has also presented a parallel plate photovoltaic amplifier connected to a potentiometer as a Resistance-Capacitance (RC) circuit power amplifier. The effect of inductance and resulting power transfer has been theoretically determined in the RC amplifier circuit. The electrical and thermal properties and measurements from a parallel plate photovoltaic amplifier were collected to analyze the unbalanced power transfer and inductance in a nonlinear RC circuit amplifier using equivalent transfer functions. The concept of Wireless Information and Power Transfer using Electromagnetic and Radio Waves of Solar Energy Spectrum is also briefly outlined.

Keywords: wireless power transfer, solar energy, energy harvesting, photovoltaics, amplifier

1. Introduction

Wireless power transfer (WPT) is defined as the transmission of electrical power without wires through various methods and technologies using time-varying electric, magnetic, or electromagnetic fields. The development of various technologies for wireless power transfer is being taken widely across the power electronics domains. There are applications which include microwaves, solar cells, lasers, and electro-magnetic waves' resonance in wireless power transfer. With wireless power transfer, the electrical devices are continuously charged without the use of power cord. The three types of wireless power transfer systems can be described by microwaves, resonance, and solar cells. From the power source to a receiver in an electrical device or gadget, microwaves can be used to send electro-magnetic radiation. The method of resonance can be applied at certain frequencies to cause an object to oscillate by electro-magnetic radiation. These oscillations can be used to transfer energy between two oscillating sources. The satellite in space with solar cells can be used to capture the solar energy and transmit this energy back to earth. This method would involve the conversion of radio waves frequencies into electrical power and electrical power into radio waves frequencies. The main purpose of the solar photovoltaic system is to distribute the collected electrical energy in various small-scale power applications wirelessly. These recent developments give technology based on how to transmit electrical power without any wires, with a small-scale by using solar energy.

Wireless power transfer (WPT) using solar energy technology is having vast applications. The ability of technology is to transfer power efficiently, safely over distance can improve gadgets and products by making them more reliable, climate and environment benign. Wireless power transfer (WPT) can be used in various applications for example in automatic wireless charging, direct wireless power supply of devices such as cellphones, loudspeakers, digital picture frames, flat screen TV's, home theater accessories etc. [1]. The power can also be transferred wirelessly through an inductive coupling as an antenna. With this wireless electricity we can charge and make wireless electricity as an input source to electronic equipment such as Handphone, MP3 Player etc. In harvesting energy, technologies of ambient solar radiation like solar photovoltaic, kinetic, thermal or electro-magnetic (EM) energy can be used to recharge the batteries. Radio frequency (RF) harvesting technologies are also popular as they are enormously available in the atmosphere. The energy converted to useful DC energy which can be used to charge electrical devices which need low power consumption.

This chapter outlines the recent developments of wireless power transfer using solar energy. The rest of the chapter contains brief history of the development of wireless power transfer. Various methods and technologies used in wireless power transfer are outlined. The State-of-the-Art of Wireless Power Transfer using Solar Energy is also described along with the literature review. The later part of the chapter contains novel concept of transmitter design of a parallel plate photovoltaic amplifier device integrated in a Building. The design of a receiver using radio waves for wireless information and power transfer is also briefly discussed. Conclusions and equations for design of a transmitter and a receiver are provided in the later part of the chapter.

1.1 History

The presence of electro-magnetic waves by devising a mathematical model is predicted by James C. Maxwell in 1864. The Poynting Vector would play an important role in quantifying the electromagnetic energy (John H. Poynting, 1884). Heinrich Hertz first succeeded in showing experimental evidence of radio waves by his spark-gap radio transmitter in 1888, which was bolstered by Maxwell's theory. The wireless power transfer was started by the prediction and evidence of the radio wave in the end of 19th century. Wireless power transfer of electrical power was pioneered by Nikola Tesla [2]. He conducted experiments on wireless power in 1891 at his "experimental station" at Colorado. A small incandescent lamp by means of a resonant circuit grounded on one end was successfully lighted by Nikola Tesla [3]. The lower end connected to the ground and the upper end free with a coil outside his laboratory. The current was induced in the three turns of wire wound around the lower end of the coil and the lamp was lighted. For trans-Atlantic wireless telephony and demonstration of wireless electrical power transfer by means of Wardenclyffe tower, which was designed by Tesla. The modern development of microwave power transmission which dominates research and development of wireless power transfer today was achieved by William C. Brown. In the early 1960s Brown invented the rectenna which directly converts microwaves to DC current. He demonstrated its ability in 1964 by powering a helicopter from the solely through microwaves.

1.2 Methods

Radio and Microwave: Typically, in the microwave range, wireless power transfer via radio waves can be made over longer distance power beaming, with shorter wavelengths of electromagnetic radiation with more directional component. To convert the microwave energy back into electricity, a rectenna may be used. Conversion efficiencies exceeding 95% have been realized with rectenna. For the transmission of energy from orbiting solar power satellites to Earth and the beaming of power to spacecraft leaving orbit, power beaming using microwaves has been considered [4].

Electromagnetic Transmission: Electromagnetic waves can also be utilized for wireless power transfer. Power beaming can be employed by converting electricity into light, such as a laser beam, then firing this beam at a receiving target, such as a solar cell on a small aircraft, power can be beamed to a single target.

Induction: For the transfer of wireless electrical power, the principle of mutual induction between two coils can be used. Electromagnetic coupling between the two coils is used to transfer the energy. The simplest example of how mutual induction works is the transformer, where there is no physical contact between the primary and the secondary coils.

Electrodynamic Induction: Resonant inductive coupling for wireless power transfer resolves the main problem associated with non-resonant inductive coupling. It has dependence of efficiency on transmission distance. The transmitter and receiver inductors are tuned to a mutual frequency and the drive current is modified from a sinusoidal to a non-sinusoidal transient waveform with the use of resonant coupling. Pulse wireless power transfer occurs over multiple cycles. Significant wireless power may be transmitted over a distance of up to a few times the size of the transmitter with this method.

Electrostatic Induction: For wireless energy transfer involving high frequency alternating current potential differences transmitted between two plates or nodes, capacitive coupling is utilized with an electric field gradient or differential capacitance between two elevated electrodes over a conducting ground plane.

1.3 Technologies

Microwave Transmitter: The frequency range of choice for transmission is achieved by means of microwaves. Presently, an efficiency of 76% is possible using current technology for microwave power transmission. The waves must be focused so that all the energy transmitted by the source is incident on the wave collection device for increasing transmission efficiency. The high cost of transmitters and the relative low efficiency of current optical and infrared devices makes higher frequencies impractical. The most common transmitters for microwaves are klystron, traveling wave tube (TWT) and magnetron. The klystron has been the DC to microwave converter of choice however it is also somewhat expensive. The TWT is far too expensive and power restrictive making it impractical for the task of power transmission. The use of magnetrons because they are cheap and efficient is investigated by many researchers. The power transmission is more lenient to frequency fluctuations than the communication systems in magnetrons frequency output but is not as precisely controllable as the klystron or TWT. An array of magnetrons to be used as the transmitter can be one of the more common R & D investigation proposal. In range of 300 W to 1 kW magnetrons are already mass produced for microwave ovens, it is one of the main advantages to using many smaller magnetrons as opposed to a few klystrons.

The generation of microwave power in the microwave power source and its output power is managed by electronic restrain circuits on the transmission side. To match the impedance between the transmitting antenna and the microwave source, a tuner is attached. Based on the direction of signal propagation by Directional Coupler, whose function is to divide the attenuated signals. The transmitting antenna emits the power uniformly through free space to the receiver antenna. An antenna receives the transmitted power and translates the microwave power to DC power on the receiving section. For setting the output impedance of a signal source equal to the rectifying circuit, both impedance matching circuit and the filter is provided. The Schottky barrier diodes which converts the received microwave power into DC power are connected in the rectifying circuit.

Use of Microwave Power Transmission in Solar Power Satellites (SPS): For transmitting power to earth stations, solar power generating satellites can be launched into space. Based on this idea, which was first proposed in 1968 based on experiments carried out in terrestrial laboratories. At high earth orbit in geosynchronous location, the SPS satellites are put in the orbits. This feature enables them to receive light almost whole year by up to 99% of the yearly time. A facility of a large rectenna array built on the Earth is for collecting the incoming microwaves. The satellite is required to be built with a retrodirective transmitter for maintaining a good lock on the rectenna. This helps in locking on to a pilot beam emanated from the ground station. Most of the research is done in the 2.4 GHz to 5.8 GHz range. Therefore, there are some spectrum regulatory issues to deal with their use. Also, the retro directive antenna system is unproven with present technology. The microwave beam could veer off target and can microwave some unsuspecting family. Therefore, this is the cause of the health concern [5].

Magnetic Resonance: In this technology, an oscillator is designed to generate the carrier signal for transmitting the power. Usually, oscillators are not intended to deliver the power, because a power amplifier is required to the oscillator for amplification of the oscillating signal. The output power to the transmission coil is transferred by the power amplifier. For receiving the transmitted power, a receiver coil is built. Since the power received at the receiver side is having an alternating current. Thus, a rectifier is needed for rectification of the AC voltage. An electric load is connected to the receiving coil to complete the circuit [6].

WiTricity: The new technology called WiTricity is based on using coupled resonant objects [1]. With the same resonant frequency, two resonant objects manage to exchange energy efficiently, while interacting weakly with extraneous off-resonant objects. The resonant nature of the witricity system guarantees the strong interaction between the sending unit and the receiving unit, while the interaction with the rest of the environment is weak. The design consists of two copper coils, each a self-resonant system. One of the coils, attached to the power source can be a solar power and is termed as the sending unit of the witricity. The irradiation of the environment with electromagnetic waves oscillates the space around it with a non-radiative magnetic field oscillating at MHz frequencies. The non-radiative field intervenes the power exchange with the receiving coil, which is built for the purpose of creating resonance with the field.

2. State-of-the-art: wireless power transfer using solar energy

Solar cells are semiconductor devices in which incident sunlight releases electric charges so they can move across the semiconductor freely and thus generate an electric field to light a bulb or power a motor. The whole phenomenon of producing an electric field of voltages and currents across the solar cell is known as the photovoltaic effect [6]. The incident light for solar cells—sunlight—is freely available and abundant. The intensity of sunlight near the surface of the earth is at the most in the range of one thousand watts per square meter known as 1 sun. The cost must be considered in calculating the cost of the electricity produced by solar cells as the area occupied by the photovoltaic modules power generating system may be relatively large. The cost per unit output is the decisive factor relative to that of alternative power sources, for acquiring, installing, and operating the photovoltaic system. This is dependent on this sole factor that determines whether the solar cells will be used to supply electricity in a given situation. Solar cells are economically competitive with alternative sources in their use in terrestrial applications. The examples of these applications include pumps, communication and refrigerated devices located in remote areas far from existing transmission and distributed power lines. The markets for solar cells are growing rapidly as the cost of power from conventional sources rises, and as the cost of solar cells reduces because of technological improvements with economies at a bigger scale manufacture.

Working of a Solar Cell: The working of a solar cell depends upon the phenomenon of photo-electricity, i.e., the liberation of electrons by light falling on a body. The application of this photo-electric phenomenon to semi-conductors such as silicon has proved to be of great use. To displace an electron from a fixed position in the material and make it move freely in the material, a vacant electron position or 'hole' is created in the material by light waves when they strike a semiconductor material with sufficient energy. If a neighboring electron leaves its site to fill the hole site, this hole acts as positive charge and can move this electron. The electron–hole pairs are differentiated by the voltage in the cell material, and this creates a current. By adding small amounts of dopants and impurities to the pure material and by joining two semiconductor materials, an intrinsic voltage may be created. The silicon becomes electron-rich and is referred toas 'n-type' silicon when impurities such as phosphorous are introduced into this silicon. Excess holes are created when impurities such as boron give rise to 'p-type' silicon. A free charge leaks across the common boundary of these n-type sand p-type silicon (one electron rich and the other electron-deficient) and becomes fixed as ions in the region near to the boundary, when these two oppositely charged semiconductors are in contact. At the interface, the fixed (but opposite) ions create an electric field that sends free electrons one way and free holes the other side. No current flows in the solar cell, when no light falls on its surface i.e., in the dark. A current will flow as long as the solar cell is illuminated which can supply electricity to an external load circuit. The current from the solar cell passes directly through the load circuit. The current generated can be changed by the power-conditioning equipment to alternating current at the voltage and current levels different from those provided by the solar cells cell. The sub-systems of the PV module system include energy-storage devices such as concentrated lenses, batteries and mirrors that focus the sunlight onto a smaller and hence less costly semiconductor solar cell. If concentration system is utilized, a tracking subsystem may be required to keep the array pointed at the sun throughout the day.

Maximum power point tracking: Maximum power point tracking (MPPT) is used to maximize the power output from wind turbines and photovoltaic (PV) solar systems. PV solar systems exist in several different configurations with a solar inverter, which is connected directly to the electrical grid. A second adaptation which is called the hybrid inverter. In this hybrid inverter, the most basic version sends power from the solar panels directly to the DC-AC and splits the power at the inverter, where a percentage of the power goes to the grid and the remainder goes to a battery bank. The third type uses a dedicated PV inverter that features the MPPT in which the inverter is not connected at all to the grid. The power flows directly into the battery bank in this configuration. The micro inverters are deployed, one for each PV panel, which are a variation on these configurations. The efficiency of solar PV system by up to 20% by the use of microinverter. The grid-connected power as well as solar PV power and branching off power for battery charging is achieved by incorporating a new MPPT algorithm that is equipped with specialty inverters which serve these three functions,

The application related to solar photovoltaic systems contains these MPPT apprehensions. A non-linear output efficiency which can be analyzed based on the I-V curve of the solar cells establishes a complex relationship between temperature and total resistance that produces across solar cells. The output of the PV cells and application of the proper load to obtain maximum power for any given environmental conditions is achieved by the purpose of the MPPT system. MPPT devices are connected into solar photovoltaic system for providing voltage or current conversion, filtering, and regulation for driving various loads, including power grids, batteries, or motors. Solar power inverters are used to convert the DC power to AC power after utilizing MPPT.

Solar Photovoltaic System Technology for Wireless Power Transfer: The solar photovoltaic panels can be installed on the façade or roofs. These solar photovoltaic panels convert the sunlight into the direct current (DC) power. The electric current is added or drawn from the electric batteries by means of installing the charge controller, which limits the rate of the current. The batteries are one of the most important parts of the solar power system. The charge controller helps in protecting the batteries from overvoltage and overcharging. This helps in increasing the life span of the batteries. From the solar photovoltaic panels, the DC power is transmitted to the inverter. In the inverter, it is converted into alternating current (AC) power.

The phase locked loop oscillator with a Power Amplifier is connected to the solar inverter. A step up/down transformer is connected to this end section. The generation of an output signal whose phase is related to the phase of the input signal is achieved by means of the phase locked loop oscillator. There is generation of a periodic signal by means of the phase locked loop oscillator. The comparison of the phase of that signal with the phase of the input periodic signal and corrects the oscillator to keep the phases matched is achieved by means of the phase detector. The power amplifier is used to achieve high amplification of the signal. The stepping up or stepping down the signal, which can be done according to the application is achieved by means of the transformer which is connected to the end section of the amplifier. In the AC line, this alternating current is then transmitted. For powering the connected load or other domestic devices, the power from these AC lines is achieved by means of wireless power transfer.

The principle of witricity can be applied into this scenario [1]. To transfer wireless power between two electromagnetic resonant objects, Witricity can be used which is based on strong coupling. This method is different from other methods like air ionization, microwaves, and induction. The witricity system consists of transmitters and receivers. These contain magnetic loop antennas critically tuned to the same frequency. Due to the operation in the electromagnetic near field, the receiving devices must be no more than the quarter-wavelengths from the transmitter. The witricity uses near field inductive coupling through magnetic fields like those found in transformers. These tuned magnetic fields generated by the primary coil can be arranged to interact actively with matched secondary windings in distant equipment. These magnetic fields are far from more-weakly with any surrounding objects or materials such as radio signals or biological tissue [6–13].

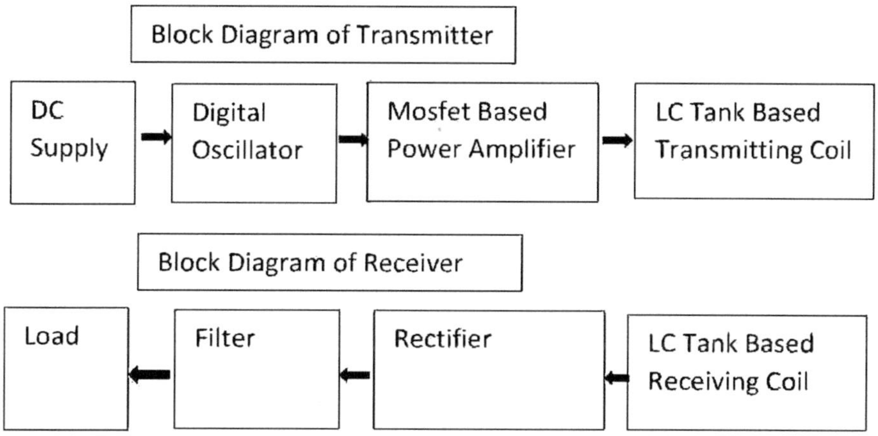

Figure 1.
Simplified block diagram for wireless power transfer using solar energy.

Working of a Transmitter: The input from mains is given to the power and frequency controller. The output of this system is given to MOSFET (metal–oxide–semiconductor field-effect transistor)/IGBT (insulated-gate bipolar transistor). The objective of using the MOSFET/IGBT is for conversion of DC power to AC power. It is also used for amplifying square wave at the gate input. The voltage transmitted to the transmitting coil generates magnetic field around it. The capacitor, which is connected to the coil in parallel helps in achieving the resonating circuit. The magnetic field get induced in the receiving coil at the point of the resonant frequency of receiving coil matches with the resonant frequency of the transmitting coil. Different values of "L" and "C" for resonant frequency are used for the matching purpose. To match the resonant frequency of the receiver and the transmitter coil, the switches to vary the time periods of the square wave by controlling the frequency at output can be used [14].

Working of a Receiver: The receiving coil comes in the range of the magnetic field of the transmitting coil. This helps in achieving the voltage across the transmitting coil, which gets induced in the receiving coil because of mutual inductance. This also helps in matching of resonance frequency at the received voltage is in AC power form. Here the AC power is converted into DC for DC load, where rectifier circuit can be used to provide constant DC at the output for driving the load. And if the load is AC power load then it can be given directly to the output.

A simplified block diagram for wireless power transfer using solar energy technology is illustrated in **Figure 1** [14].

3. Literature review: wireless power transfer (WPT) using solar energy

Only few relevant papers which highlight solar energy based wireless power transfer are briefly discussed here. Zambari et al., investigated the development of wireless energy transfer module for solar energy harvesting [11]. They studied the module of wireless energy transfer (WET) for interaction with the ambient solar energy. The main objective was to distribute the collected electrical energy from a solar panel module to in house loads appliances wirelessly. The investigations were carried out on the solar panel module with 240 W, 30 V, Poly Crystalline Silicon Photovoltaic solar panel. The design of the WET module was based on magnetic resonance technology. This technology uses two sub-unit modules development;

driving circuit and two coils mutually inducted to transfer energy in a suitable resonant frequency. With the advantage of nearly 99% efficiency theoretically, class-D RF power amplifier was used as the driving circuit for transmitted coil switching [11].

Fareq et al., studied the wireless power transfer by using solar energy [12]. They developed the project based on electrical power without any wires, with a small-scale by using solar energy. The power is transferred wirelessly through an inductive coupling as an antenna. The experiments were conducted and the wireless power transfer can be transfer energy up to 10 cm. with efficiency 0–10 cm; 98.87% -40% [12]. Ojha et al., investigated solar energy based wireless power transfer [13]. They reviewed on wireless power transfer (WPT) using renewable source i.e. solar energy. The principle behind WPT used was inductive coupling wherein an electric field is generated thus transmitting power from transmitter to receiver. The paper has highlighted the important use of components like a solar panel, rechargeable battery, booster circuitry, and load. Wireless transmission of power to work up a load was highlighted in the paper [13]. Lakshmi M. K., et al. investigated wireless power transmission through solar power generation [14]. The phenomenon of transfer power using a renewable source, without using wired medium. This paper mainly focused on combining both wireless and solar technologies together with use of the principle through coupled resonant objects for the transferring electricity. The overall goal of this paper is to design and implement a clean power generation and wireless power transmission system which can be used as a standard means for charging any electronic gadget [14].

Maqsood et al., investigated wireless power transmission using solar based power satellite technology [15]. The wireless electricity (Power) transmission (WET) was focal point of their research and they presented the concept of transmitting power wirelessly to reduce transmission and distribution losses. The wired distribution losses are 70–75% efficient. The paper also highlighted the benefits of using WET technology specially by using Solar based Power satellites (SBPS) and also focused on how we make electric system cost effective, optimized and well organized [15]. Keerthana et al., investigated Wireless Power Transfer Using Rectenna [16]. The Radio frequency (RF) harvesting technologies were highlighted in the paper. The RF harvesting technologies receive and convert the useful DC energy and can further be used to charge electrical devices which need low power consumption. The paper investigated a microstrip square patch antenna operating at 2.45 GHz. It was fabricated on a low-cost FR4 substrate having a dielectric constant of 4.4 with a thickness of about 1.2 mm. The L-shaped matching network was designed for maximum power transfer between the antenna and the rectifier. The HSMS-2850 zero bias Schottky diode was used as a rectifier. The RF-DC rectification was done with an efficiency of 42.8% at -7 dBm at 2.45 GHz [16].

4. Transmitter design: a parallel plate photovoltaic amplifier device integrated in a building

A parallel plate photovoltaic device connected to a potentiometer is analyzed as a star connected 3-Phase Resistance-Capacitance (RC) circuit amplifier. The effect of inductance and resulting power transfer has been determined in the RC circuit amplifier constituting of a parallel plate photovoltaic device. The analysis has also discussed from the electrodynamics point of view, power transfer and effect of induction losses in a 3-Phase RC circuit amplifier constituting of a parallel plate photovoltaic device. The theory of the sinusoidal steady-state response was applied in performing the analysis of the circuit, because of the advantage of representing a periodic function in terms of a sinusoidal exponential function. The full-scale experimental setup for a parallel plate photovoltaic device connected to a potentiometer was installed in an outdoor room facility located at Concordia University, Montréal, Canada [17–34]. The analysis has been performed on the basis of the accepted unified theory for stresses and oscillations, as proposed by the author [27]. The experimental setup is illustrated in **Figure 2** [17]. A pair of glass coated photovoltaic (PV) modules forming a parallel plate duct with a plywood board and connected to a potentiometer was used to build an amplifier. A wire-wound variable resistor with resistance up to 50 Ω was a wire-wound circular coil with a sliding knob contact acted as a potentiometer for the circuit [17]. This potentiometer was used to vary electrical resistance across connected PV modules without interrupting the current. A star connected RC circuit amplifier with a parallel plate photovoltaic device connected to a potentiometer was built (**Figure 3**).

A Photovoltaic (PV) Device connected to a potentiometer: A parallel plate photovoltaic device connected to a potentiometer was used for establishing characteristics of this PV device by varying electrical resistance with rotation of knob of

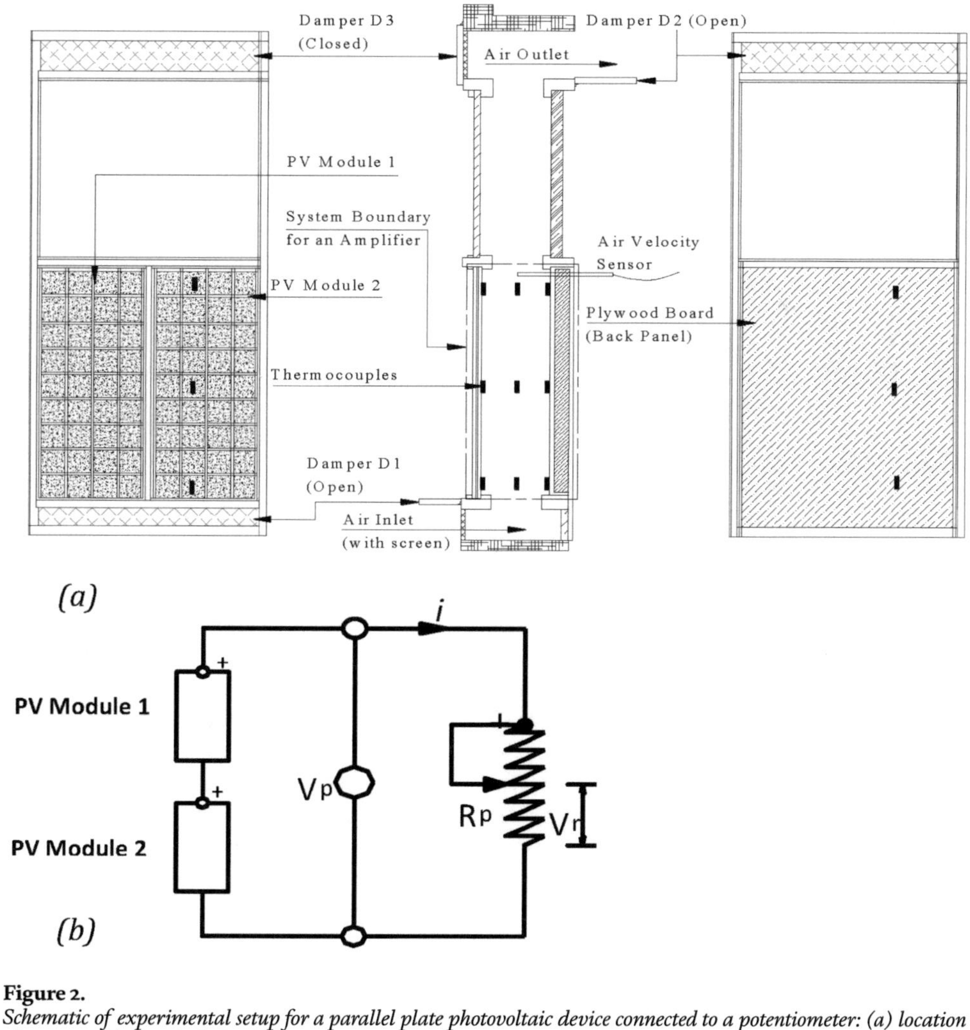

Figure 2.
Schematic of experimental setup for a parallel plate photovoltaic device connected to a potentiometer: (a) location of sensors; (b) electrical circuit diagram.

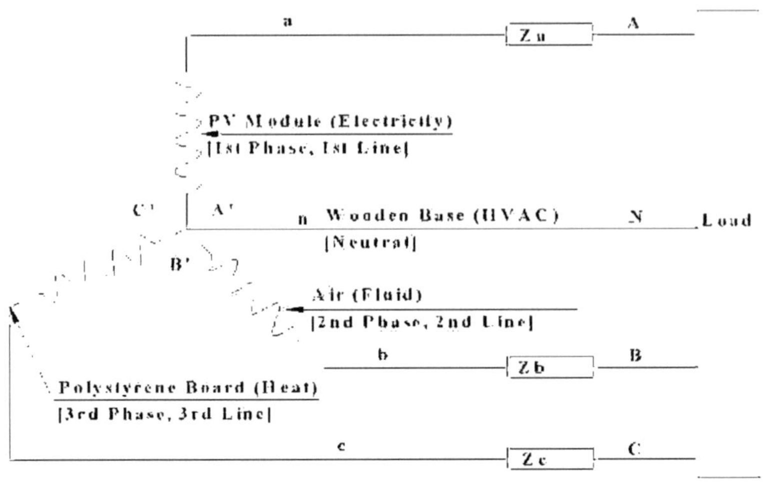

Figure 3.
A star connected 3-phase generation using a parallel plate photovoltaic device.

Rotation	Volts	Amps	Watts	Rotation	Volts	Amps	Watts	Rotation	Volts	Amps	Watts
240°	18.7	—	—	83°	16.3	0.935	15.23	30°	9.7	1.577	15.21
239°	16.5	0.331	5.461	75°	16.0	1.014	16.26	27°	9.0	1.587	14.33
201°	17.4	0.414	7.195	69°	15.8	1.100	17.38	21°	7.1	1.583	11.24
185°	17.5	0.454	7.940	64°	15.5	1.165	18.04	18°	6.2	1.573	9.831
162°	17.3	0.513	8.885	55°	15.0	1.302	19.53	17°	5.7	1.578	9.026
150°	17.18	0.550	9.449	50°	14.5	1.386	20.05	12°	3.9	1.567	6.257
142°	17.19	0.582	10.00	43°	13.2	1.503	19.79	10°	3.2	1.553	4.840
128°	17.1	0.640	10.93	42°	13.1	1.493	19.49	1.5°	0.5	1.593	0.807
107°	16.8	0.750	12.51	37°	11.9	1.536	18.26	1°	0.3	1.59	0.426
89°	16.4	0.884	14.45	32°	10.5	1.567	16.42	0°	—	1.643	—

Table 1.
Sample electrical measurement results with varying resistance of potentiometer.

a potentiometer [17]. For determining electric power output with a series electrical circuit connection of a pair of vertically inclined PV modules installed on a wooden frame, the current–voltage measurements were obtained. The electrical measurement results of currents, voltages and power with varying electrical resistance of potentiometer are presented in **Table 1**. The results of the power output from a potentiometer with rotation of circular knob are illustrated in **Figure 4**. The phenomenon of photovoltaic amplification has been observed from the graph of **Figure 4**. The gain in steady state electrical for a photovoltaic device is a factor of its volume or resistance. This operational characteristic is similar to the operation of a loudspeaker.

5. Electrical parameters for the RC circuit amplifier

Capacitance: The capacitance of a parallel plate photovoltaic device with air as a dielectric medium was calculated to be 91.2 picofarads.

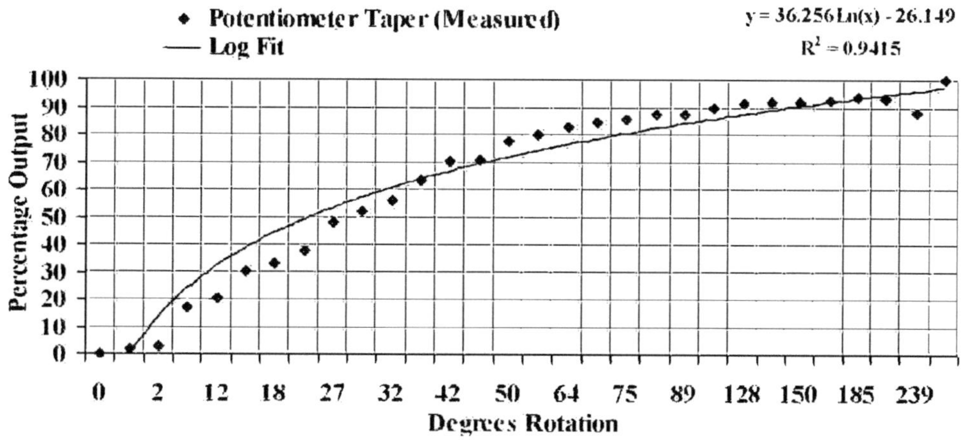

Figure 4.
Potentiometer taper as a function of percentage voltage output.

Resistance: The electrical resistance of various components were calculated as: glass coated PV modules were approximated as 5.3 kΩ, air was approximated as 1200 MΩ, and plywood board was approximated as 26.5 Tera Ω. The total equivalent electrical resistance of a parallel plate photovoltaic device was approximated as: 5.3 kΩ.

Time Constant: The time constant, which is product of resistance and capacitance, was calculated to be: 0.5 microseconds. The frequency with this time constant was calculated to be 2 MHz.

Capacitive Reactance: The capacitive reactance was calculated to be: 872.5 Ω.

Impedance: The impedance of the circuit was calculated to be: 5.4 kΩ.

The Phase angle θ*:* The phase angle between capacitance and reactance was calculated to be 9°.

The Phasor representation: $Z = 5.300 - j\,0.8725 = 5.4\ k\Omega \angle{-9°}$.

Capacitive Heating: The joule law gives instantaneous power absorbed by the capacitive impedance and is converted to heat. The heat capacities under critical operation of buoyancy-induced hybrid ventilation were calculated to be 59.6 kJ, 0.755 kJ and 510.7 kJ for PV module, air and plywood board respectively [28]. The total average value of joule heating for the parallel plate photovoltaic device was calculated to be 571 kJ.

Induction Losses: The induction losses due to thermal storage effect in the parallel plate photovoltaic device was calculated to be 15.9 KJ [28].

Power Factor: The power factor was calculated to be cos θ = 0.911 lag.

Current function **(i2(t))**: Using the current function, $i^2(t) = I_m^2 \sin^2(\omega t + \theta)$, the effective (root mean square) value of current was calculated to be 10.4 amps and maximum value of current was calculated to be 14.71 amps.

Voltage function: The voltage function is defined as per the sine wave: $v = V_m \sin(\omega t)$. The effective value of the voltage was calculated to be 60.4 volts and maximum value of the voltage was calculated to be 85.42 volts.

Power function: The instantaneous power is given by the expression:

$$p(t) = \frac{V_m I_m}{2}\cos(\theta) - \frac{V_m I_m}{2}\cos(2\omega t - \theta) \quad (1)$$

The Plots: The time diagram for current, voltage is plotted in **Figure 5(a)**. The time diagram for power is plotted in **Figure 5(b)**.

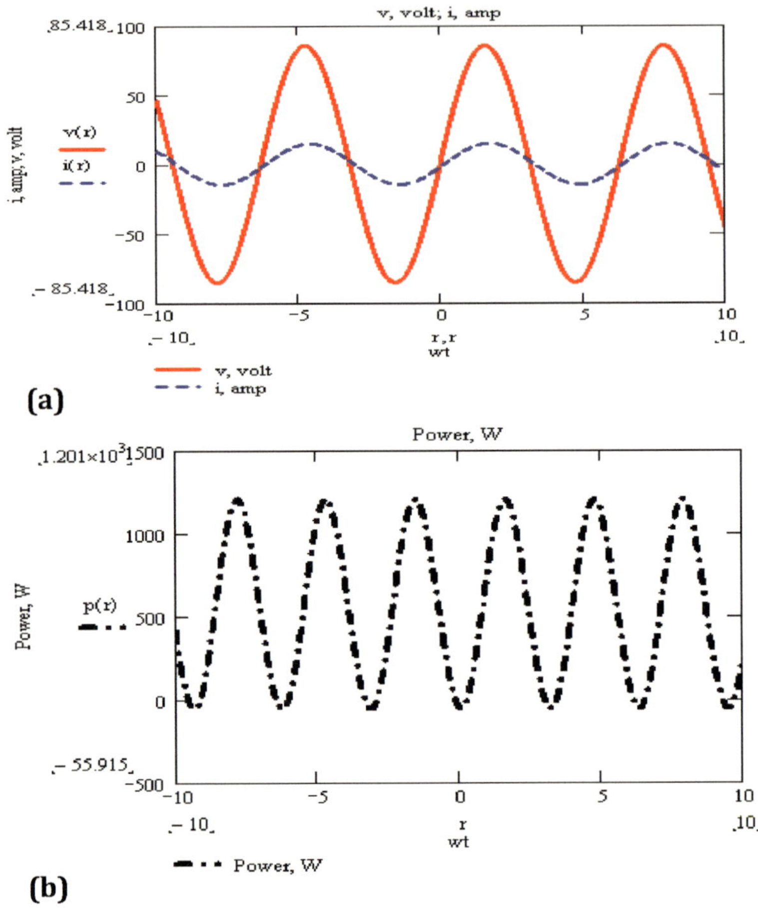

Figure 5.
Time diagrams: (a) voltage and current; (b) power in the RC circuit amplifier.

5.1 Discussions on power transfer and effects of inductance

Capacitive Reactance and Resistance in Series: The losses that appear in capacitive circuits are lumped in a resistor connected in series with the capacitor.

Capacitance and Resistance in parallel: When a sine waveform voltage is applied across a capacitor and a charging current of sine waveform is transmitted across the circuit. The alternating current source, which is a sine waveform voltage, if applied at a uniform rate, is responsible sinusoidal response of the charging current in the capacitor. The motion of the charging current is transmitted through the capacitor, corresponds to the electron flow in the wires connecting the capacitor to the alternating current source. The alternating current source is responsible for development of the electric stress in the dielectric between the plates of the capacitor. Electron flow does not occur through the capacitor. The electrons flow around the capacitor circuit in one cycle, which causes a negative charge to build up on one place, and a corresponding positive charge on the other, and the next cycle causes a reversal of the polarity of the charges on the plates. Thus, the effective impedance which the capacitor offers to the flow of alternating current can be relatively low while the insulation resistance which the dielectric offers to the flow of direct current is extremely high.

Power Transfer: With no voltage or charge, the electrons in the dielectric between the capacitor plates rotate around their respective nuclei in normally circular orbits.

When the capacitor receives a charge the positive plate (PV Module) repels the positive nuclei and at the same time the electrons in the dielectric are strained toward the positive plate and repelled away from the negative plate (Plywood Board). This distorts the orbits of the electrons in the direction of the positive charge. During the time the electrons are changing from normal to the strained position there is a movement of electrons in the direction of the positive charge. The movement constitutes the displacement current in the dielectric. When the polarity of the plate reverses, the electron strain is reversed. If a sine-wave voltage is applied across the capacitor plates the electrons will oscillate back and forth in a direction parallel to the electrostatic lines of force. Displacement current, is a result of the movement of bound electrons, whereas conduction current represents the movement of free electrons.

The **Figure 5(b)** shows that the instantaneous power is negative whenever the voltage and current are of opposite sign. However, as has been illustrated in the **Figure 5(b)** that positive area of p(t) energy exceeds the negative area. Therefore, the average power is finite. Since the angle, θ, is small between current and voltage, the negative area of p(t) energy become very small. During the first quarter cycle (from 0° to 90°) the applied voltage rises from slightly negative value to a maximum and the capacitor is receiving a charge. The power curve is positive during this period and represents energy stored in the capacitor. From 90° to 180°, the applied voltage is falling from maximum to slightly negative value and the capacitor is discharging. The corresponding power curve is negative and represents energy returned to the circuit during this interval. The third quarter cycle represents a period of charging the capacitor and the fourth quarter represents a discharge period.

Induction Losses: The induction losses due to thermal storage amount to 1.5% in comparison to the capacitive heating [28, 29]. When a circuit containing a coil or source of energy is energized with direct current, the coil's effect in the circuit is evident only when the circuit is energized, or when it is de-energized. However, when the inductive circuit is supplied with alternating current, the induction losses are continuous and much greater than when it was supplied with direct current. For equal applied voltages, the current through the circuit is less when alternating current, is applied than when direct current is applied. The alternating current is accompanied by an alternating magnetic field around the area of the source of energy, which cuts through the area of the source of energy in the circuit. Most of the applied voltage appears across inductance, L, with little remaining for the load in the circuit. In a circuit possessing inductance only, the true power is zero. The current lags the applied voltage by 90°. The areas of induction losses above the X axis represent positive energy and the areas below the X axis represent negative energy [28–34].

6. Development of a receiver using radio waves for wireless information and power transfer

The focus of the current research is to expedite the efforts for development of a receiver using radio waves for wireless information and power transfer using solar energy spectrum. Liang Liu et al. investigated transmit beamforming for simultaneous wireless information and power transfer using radio frequency (RF) transmission [35]. It is essential to have Radio frequency (RF) transmission enabled wireless power transfer (WPT) to power energy-restricted wireless systems (e.g., sensor networks), where dedicated energy transmitters are deployed to broadcast RF signals to charge low-power electric devices (e.g., sensors and RF identification

(RFID) tags), as it is a cost-effective solution. Radio Frequency (RF)-based wireless power transfer (WPT) can provide continuous and controllable power supply, and thus is applicable to more energy-demanding applications [35]. Radio frequency (RF) signals have been widely used in wireless communications as the carrier for wireless information transfer (WIT) for several decades.

A query thus arises that whether we can utilize RF signals more efficiently for both WPT and WIT at the same time with a new technique called simultaneous wireless information and power transfer (SWIPT) [35]. The SWIPT is developed by considering a single-antenna point-to-point channel, where the trade-off between the achievable rate for WIT and the received energy for WPT is investigated that the single-antenna receiver can utilize the same received RF signals for both information decoding (ID) and energy harvesting (EH) at the same time without any loss. However, this assumption is difficult to realize in practice since existing information receivers (IRs) and energy receivers (ERs) are separately designed with distinct circuit structures, and as a result, each of them cannot be used to decode information as well as harvest energy at the same time. The two basic receiver structures have been widely adopted in the literature [35].

Time-Switching receiver (TS) switches between an information decoder and an energy harvester over time. This technique is the simplest way to implement SWIPT by using off-the-shelf commercially available circuits for information decoding (ID) and energy harvesting (EH), respectively. It is crucial to determine their operation modes (ID or EH) over time for TS receiver. This is based on their communication and energy requirements, as well as the channel conditions [36]. Power-Splitting receiver (PS) splits its received signal into two portions with one for information decoding (ID) and the other for energy harvesting (EH). In this technique, it is important to determine the power splitting ratio at each antenna to balance the rate-energy trade-off between the information decoding (ID) and energy harvesting (EH) receivers. Note that time-switching and power-splitting receiver can be regarded as a special and low-complexity realization of power-splitting receiver with only binary (0 or 1) power splitting ratio at each receiving antenna. They are implemented by different hardware circuits (time switcher versus power splitter) in practice [36].

There are miscellaneous issues investigated by many researchers on wireless power transfer. A. M. Azman et al., investigated superimposition technique in wireless power transfer for enhancing the distance of transmission of the transmission coil [37]. This technique resulted in incrementation of the distance by up to 2 times compared to the system without superimposed technique. Yunfei Chen et al., investigated interference analysis in wireless power transfer [38]. They studied the co-channel interference (CCI) generated by wireless power transfer. They considered the effect on information delivery for three widely used setups of simultaneous wireless information and power transfer (SWIPT), hybrid access point (HAP) and power beacon (PB). In the book on Wireless Power Transfer edited by Johnson I. Agbinya, various innovative techniques for design of Optimal Wireless Power Transfer Systems are discussed [39]. The authors present new methods of delivering flux efficiently using the inductance-based transmitter to an inductance-based receiver by using either flux concentrator or separator. The flux coupling coefficient can be increased by the concentrator. This leads to increased flux delivered to the receiver by a large order of magnitude. Whereas the separator helps in reducing the crosstalk between two identical types of nodes and also leads to significant increase in power delivery. In another paper, Zhen Zhang et al., investigated energy encryption for wireless power transfer [40]. They studied the improved security performance of wirelessly transferred energy as an attempt to switch off other unauthorized energy transmission channels and enhancing security of energy transmission.

7. Conclusions

This chapter has presented brief outline of the state-of-the-art and developments in wireless power transfer using solar energy. The harvesting technologies of ambient solar radiation like solar photovoltaic, kinetic, thermal or electro-magnetic (EM) energy can be used to recharge the batteries and power various electronic gadgets. Brief on Radio frequency (RF) harvesting technologies is also presented. The energy converted to useful DC energy which can be used to charge electrical devices which need low power consumption. The chapter has also presented analysis of the parallel plate photovoltaic amplifier connected to a potentiometer as a Resistance-Capacitance (RC) circuit power amplifier. The effect of inductance and resulting power transfer was theoretically determined in the RC amplifier circuit. The electrical and thermal properties and measurements from a parallel plate photovoltaic amplifier were collected to analyze the unbalanced power transfer and inductance in a nonlinear RC circuit amplifier using equivalent transfer functions. The concept of Wireless Information and Power Transfer using Electromagnetic and Radio Waves of Solar Energy Spectrum is also briefly outlined. The chapter has also presented miscellaneous issues pertaining to wireless power transfer such as superimposition technique, interference, and security issues. Appendix has presented Equations for transmitter and receiver using mutual inductance of the magnetic resonance between transmitter and receiver.

Appendix: Equations for design of a transmitter and a receiver

A transmitting antenna is surrounded by an electromagnetic field. This electromagnetic field is divided into two separate regions-the reactive near field and the radiating field. The energy is stored in the transmitting coil before it propagates as electromagnetic waves to the receiving coil [41].

The magnetic field experience between transmitter or receiver is called mutual inductance, which can be predicted through:

$$\mathrm{M} = \mathrm{L}_1\,\mathrm{L}_2 \tag{A1}$$

Where, L_1 is inductance of transmitter coil and L_2 is inductance of receiver coil.

For circular loop coil the inductance can be calculated by using the following formula [42]:

$$L = N^2 \mu_o \mu_r \frac{D}{2}\left(\left(\ln\frac{8D}{d}\right) - 2\right) \tag{A2}$$

Where

N – Number of turns of the coil

$\mu = 4\pi \times 10^{-7}$ permeability of vacuum, (H/m)

D – Diameter of loop coil (m).

d – Diameter of conductor cross-section (m).

The coil inductance (L) and optimal resonance frequency is determined based on operating frequency that has been used in the system, the capacitance C can be calculated by:

$$f_r = 1/2\pi\left(1/\sqrt{LC}\right) \tag{A3}$$

Quality Factor, *Q* characterize the energy decay in an antenna coil which is inversely proportional with the energy loss in antenna coil before transfer to the receiving coil. The factor Q of coil can be determined using Quality factor:

$$Q = \omega L/\left(R_{ac} + R_{rad}\right) \tag{A4}$$

Where, R_{ac} is AC resistivity; R_{rad} is radiation resistivity. The quality factor *Q* can have values ranging from 0 to infinity. It is difficult to obtain the values of Q far above 1000 for antenna coils in actual practice [43]. A high-*Q* antenna coil can be defined with *Q* greater than 100. These two coupling antenna coils should have Q greater than100 for each of the coils for transmission of the energy wirelessly [44]. The efficiency of the transfer system is very low for the antenna coils which have Q between 100 and 200 [45]. For obtaining a high efficiency of the wireless power transfer system, a high factor *Q* which approximates to 1000 is preferred for design purposes [46].

Author details

Himanshu Dehra
Monarchy of Concordia, Wellstar Beacon Labs, Faridabad, Haryana, India

*Address all correspondence to: anshu_dehra@hotmail.com

References

[1] Sagolsem Kripachariya Singh, T. S. Hasarmani, and R. M. Holmukhe, Wireless Transmission of Electrical Power Overview of Recent Research & Development, *International Journal of Computer and Electrical Engineering, Vol.4, No.2, April 2012.*

[2] Nikola Tesla, My Inventions, Ben Johnston, Ed., Austin, HartBrothers, p. 91,1982

[3] Nikola Tesla, "The Transmission of Electrical Energy Without Wires as a Means for Furthering Peace," Electrical World and Engineer. Jan. 7, p. 21, 1905.

[4] G. E. Maryniak, "Status of international experimentation in wireless power Transmission," Solar energy, vol. 56, Issue 1, 1996, Pages 87-91.

[5] R. B. Erb, "International cooperation for the acquisition of space based energy," Solar Energy, vol. 56, Issue. 1, 1996, Pages 79-85.

[6] Megha Pancholi, Nilam Panchal, Dhirendra Rana, Wireless Power Transfer by Incorporation of Solar Energy, International Research Journal of Engineering and Technology (IRJET) e-ISSN: 2395-0056 Volume: 05 Issue: 04 | Apr-2018 www.irjet.net p-ISSN: 2395-0072, Page 4055

[7] Wenzhen Fu, Bo Zhang, Dongyuan Qiu, "Study on Frequency-tracking Wireless Power Transfer System by Resonant Coupling," Power Electronics and Motion Control Conference, pp. 2658-2663, May 2009.

[8] B. Renil Randy, M.Hariharan, R. Arasa Kumar "Secured wireless powerTransmission using radio frequencySignal" DOI: 10.5121/ijist.2014.4315

[9] Xiaolin Mou," Wireless Power Transfer: Survey and Roadmap *Student Member", IEEE* and Hongjian Sun, *Member, IEEE* The research leading to these results has received funding from the European Commission's Horizon 2020 Framework Programme (H2020/2014-2020) under grant agreement No. 646470, SmarterEMC2 Project,18 feb 2011

[10] Wenzhen Fu, Bo Zhang, Dongyuan Qiu. *Study on Frequency-tracking Wireless Power Transfer System by Resonant Coupling.* College of Electric Engineering South China University of Technology Guangzhou, P. R. China. 510641.

[11] Izzul Fahmi Zambari, Chiah Yi Hui, Ramizi Mohamed, Development of Wireless Energy Transfer Module for Solar Energy Harvesting, The 4th International Conference on Electrical Engineering and Informatics (ICEEI 2013),Procedia Technology 11 (2013) 882-894, doi: 10.1016/j.protcy.2013.12.272

[12] M. Fareq, M. Fitra, M. Irwanto, Syafruddin. HS, N. Gomesh, M. Rozailan, M. Arinal, Y.M. Irwan, J. Zarinatul, Wireless Power Transfer by Using Solar Energy, TELKOMNIKA, Vol.12, No.3, September 2014, pp. 519~524, ISSN: 1693-6930, DOI: 10.12928/TELKOMNIKA.v12i3.93

[13] Shraddha Ojha, Tanuja Panda, Shweta Parab, Jyoti Dange, Renewable Energy Based Wireless Power Transfer, International Journal of Engineering Research & Technology (IJERT), ISSN: 2278-0181, Published by, www.ijert.org, ICIATE - 2017 Conference Proceedings, Volume 5, Issue 01, Special Issue – 2017

[14] Lakshmi M.K, Reenu Varghese, WIRELESS POWER TRANSMISSION THROUGH SOLAR POWER GENERATION, International Journal

of Scientific & Engineering Research, Volume 7, Issue 4, April-2016, pp. 655-661 ISSN 2229-5518 IJSER © 2016 http://www.ijser.org

[15] M Maqsood and M Nauman Nasir, Wireless electricity (Power) transmission using solar based power satellite technology, 6th Vacuum and Surface Sciences Conference of Asia and Australia (VASSCAA-6) IOP Publishing Journal of Physics: Conference Series 439 (2013) 012046 doi:10.1088/1742-6596/439/1/012046

[16] S. Keerthana, C.Uthranarayan, V. Rajalakshmi, A. Ramya, Wireless Power Transfer Using Rectenna, International Conference on Physics and Photonics Processes in Nano Sciences International Conference on Physics and Photonics Processes in Nano Sciences, Journal of Physics: Conference Series Journal of Physics: Conference Series 1362 (2019) 012037 IOP Publishing doi:10.1088/1742-6596/1362/1/012037

[17] H. Dehra, (2017). An Investigation on Energy Performance Assessment of a Photovoltaic Solar Wall under Buoyancy-induced and Fan-assisted Ventilation System, Applied Energy, Volume 191, 1 April 2017, pp. 55-74.

[18] H. Dehra, (2018). Acoustic Signal Processing and Noise Characterization Theory via Energy Conversion in a PV Solar Wall Device with Ventilation through a Room, Advances in Science, Technology and Engineering Systems Journal, Vol. 3, No. 4, 2018, pp. 130-172. https://doi.org/10.25046/aj030414

[19] H. Dehra, (2019a). Acoustic Filters for Sensors and Transducers: Energy Policy Instrument for Monitoring and Evaluating Holy Places and Their Habitants, in book 'Energy Policy' edited by Dr. Tolga Taner, https://doi.org/10.5772/intechopen.81949

[20] H. Dehra, (2019b). Principles of Energy Conversion and Noise Characterization in Air Ventilation Ducts exposed to Solar Radiation, Applied Energy, 242C, 15 May 2019, pp. 1320-1345. https://doi.org/10.1016/j.apenergy.2019.03.013

[21] H. Dehra, (2019c). Integrated Acoustic and Thermo-Fluid Insulation Modeling of an Airflow Window with a Photovoltaic Solar Wall, Building Simulation 2019 (Session: Building Acoustics), Rome, Italy, on Sep 2-4, 2019 (http://buildingsimulation2019.org/), IBPSA.

[22] H. Dehra, (2019d). Noise Calculation Charts and Indoor Environmental Quality for Evaluating Industrial Indoor Environment and Health, in book 'Indoor Environment and Health' edited by Dr. Orhan Korhan, InTech Publication, IntechOpen, DOI: 10.5772/intechopen.84993, ISBN 978-1-78984-374-3, 2019, (25 pages) (Available from: https://www.intechopen.com/chapter/pdf-download/67373).

[23] H. Dehra, (2019e). Solar Energy Conversion and Noise Characterization in Photovoltaic Devices with Ventilation, invited chapter in book, "Recent Developments in Photovoltaic Materials and Devices", ISBN 978-953-51-6690-0, edited by Dr. Natarajan Prabaharan, Dr. Marc A. Rosen and Dr. Pietro Elia Campana, IntechOpen, Chapter 1, pp. 1-20, 2019, DOI: 10.5772/intechopen.79706. (Available from: https://www.intechopen.com/chapter/pdf-download/62929).

[24] H. Dehra, A Two Dimensional Thermal Network Model for a Photovoltaic Solar Wall, Solar Energy, Vol. 83, Issue 11, November 2009, pp. 1933-1942 (10 pages), Elsevier (http://www.sciencedirect.com/science/article/pii/S0038092X08001862)

[25] H. Dehra, "Noise Transmission Losses in Integrated Acoustic and Thermo-Fluid Insulation Panels" in Book "Noise and Environment"

(Edited by Dr. Daniela Siano) https://www.intechopen.com/online-first/noise-transmission-losses-in-integrated-acoustic-and-thermo-fluid-insulation-panels, 2020, 29p.

[26] H. Dehra, "A Treatise on Solar Energy Acoustics", Book Publisher International, eBook/Short Monograph, available online at: http://www.bookpi.org/bookstore/product/a-treatise-on-solar-energy-acoustics/(33 pages).

[27] H. Dehra, (2007). A Unified Theory for Stresses and Oscillations, *Proceedings from CAA Conf. Montréal. 2007*, Canadian Acoustics. September 2007. Vol. 35. No. 3, pp. 132-133.

[28] H. Dehra, The Effect of Heat and Thermal Storage Capacities of Photovoltaic Duct Wall on Co-Generation of Electric and Thermal Power, orally presented in the proceedings of American Institute of Chemical Engineers - AIChE 2007 Spring Meeting, Houston, Texas, USA, April 22-26, 2007, session 36a (9 pages) (https://www3.aiche.org/proceedings/Abstract.aspx?PaperID=79681).

[29] H. Dehra, Power Transfer and Inductance in a Star Connected 3-phase RC Circuit Amplifier, *Proc. AIChE 2008 Spring Meeting,* New Orleans, LA, USA, April 6-10, 2008, session 96a, 7p.

[30] H. Dehra, A Multi-Parametric PV Solar Wall Device, *Proceedings from IEEE International Conference on Power, Control, Signals and Instrumentation Engineering (ICPCSI-2017)*, Chennai, India on Sep 21-22, 2017, pp. 392-401.

[31] H. Dehra, A Theory of Acoustics in Solar Energy, Natural Resources, March 2013, Vol. 4, Issue 1A (Solar Energy), pp. 116-120 (5 pages), Scientific Research, DOI: 10.4236/nr.2013.41A014 (http://www.scirp.org/journal/PaperInformation.aspx?PaperID=29361).

[32] H. Dehra, A Novel Theory of Psychoacoustics on Noise Sources, Noise Measurements and Noise Filters, *INCE Proc. NoiseCon16 Conf.*, Providence, Rhode Island, USA, 13-15 June, 2016, pp. 933-942.

[33] H. Dehra, A Paradigm of Noise Interference in a Wave, *Internoise-2018, 47th International Congress and Exposition on Noise Control Engineering*, Chicago, Illinois, USA on Aug 26-29, pp. 451-462.

[34] H. Dehra, Characterization of Noise in Power Systems, *Proceedings from IEEE International Conference on Power Energy, Environment & Intelligent Control (PEEIC2018)*, Greater Noida, India on April 13-14, 2018, pp. 320-329.

[35] Liang Liu, ... Rui Zhang, in, Transmit beamforming for simultaneous wireless information and power transfer Academic Press Library in Signal Processing, Volume 7, 2018

[36] T D P Perera, D N K Jayakody, S De, and M A Ivanov, A Survey on Simultaneous Wireless Information and Power Transfer, International Conference on Information Technologies in Business and Industry 2016, IOP Conf. Series: Journal of Physics: Conf. Series 803 (2017) 012113 doi:10.1088/1742-6596/803/1/012113

[37] W. K. M. Razali, A. Abu Bakar, F. R. Hashim, A. M. Azman, I. H. Hamzah and Z. I. Rizman, Superimposed Technique in Wireless Power Transfer, . J Fundam Appl Sci. 2018, 10(6S), 1069-1079; doi: http://dx.doi.org/10.4314/jfas.v10i6s.67

[38] Yunfei Chen, Daniel B. da Costa, and Haiyang Ding, Interference Analysis in Wireless Power Transfer, IEEE COMMUNICATIONS LETTERS, VOL. 21, NO. 10, OCTOBER 2017, pp.2318-2321.

[39] Johnson I. Agbinya Techniques for Optimal Wireless Power Transfer

Systems, in Book: Wireless Power Transfer, 2nd Edition, River Publishers, PWDelft, The Netherlands, 2016.

[40] Zhen Zhang, K. T. Chau, Chun Qiu, and Chunhua Liu, Energy Encryption for Wireless Power Transfer, IEEE TRANSACTIONS ON POWER ELECTRONICS, VOL. 30, NO. 9, SEPTEMBER 2015, pp. 5237-5246.

[41] R. C. Johnson, Antenna Engineering Handbook, 3rd ed. New York: McGraw-Hill, 1993.

[42] J. O. Mur-Mirinda, G. Fanti, Y. F. Feng, K. Omanakuttan, R. Ongie, A. Setjoadi, and N. Sharpe, "Wireless Power Transfer using weakly coupled magnetostatic resonators," 2010 IEEE Energy Conversion Congress and Exposition, Atlanta, GA, 2010, pp. 4179-4186, doi: 10.1109/ECCE.2010.5617728.

[43] E. Waffenschmidt, "Wireless Power for Mobile Device," *Telecommunications Energy Conference (INTELEC)*, pp 1-9, Oct 2011.

[44] K. L. Hall, M. P. Kesler, K. J. Kulikowski, and A. J. Campanella, "Wireless Energy Transfer Converters," US 2010/0264747 A1, pp1-78.

[45] S. A. El-Hamamsy, "Design of high efficiency RF Class-D power amplifier," *IEEE Trans. Powe Electronics*, vol. 9, no.3, May 1994.

[46] M. K. Kazimierczuk, RF Power Amplifier, A John Wiley and Sons, Ltd., Publication, pp109-178, 2008.

Chapter 3

Wireless Power Charging in Electrical Vehicles

Nassim Iqteit, Khalid Yahya and Sajjad Ahmad Khan

Abstract

Wireless Power Transfer (WPT) technology can transfer electrical energy from a transmitter to a receiver wirelessly. Due to its many advantages, WPT technology is a more adequate and suitable solution for many industrial applications compared to the power transfer by wires. Using WPT technology will reduce the annoyance of wires, improve the power transfer mechanisms. Recently, the WPT gain enormous attention to charging the on-board batteries of the Electric Vehicle (EV). Several well-known car manufacturing companies start efforts to adopt WPT technology and enhance its features. Therefore, WPT can be achieved through the affordable inductive coupling between two coils named a transmitter and a receiver coil. In EV charging applications, transmitter coils are located underneath the road, and receiver coils are installed in the EV. The inductive WPT of resonant type is generally applied to medium-high power transfer applications like EV charging because it achieves better energy efficiency. In this chapter, various WPT technologies are discussed and tested in EV wireless charging applications. Furthermore, extensive information is given to developing an advanced WPT technology that can transfer maximum power by achieving maximum efficiency.

Keywords: Wireless Power Transfer (WPT), Electric Vehicle (EV), Energy Efficiency

1. Introduction

A Wireless Power Transfer (WPT) is one of the promising technologies used to transfer electric energy from a transmitter to a receiver wirelessly. WPT is an attractive solution for many industrial applications due to its enormous benefits over wired connections. The advantages include the no hassle of carrying wires, easy charging, and smoot of power transmission even in unfavorable environmental circumstances.

The idea of wireless power transfer (WPT) was first introduced at the end of the 19th century by Nicola Tesla. He manufactured a wireless lighting bulb that was used to receive electrical charge wireless [1]. Tesla used two metal plates that were closely placed to each other. A high-frequency Alternative Current (AC) potentials were passed between these two plates, and the bulb powered ON. However, some of the issues appeared while using WPT technology. One of the main issues is that the minimum power density and low transfer efficiencies affect when the distances increase. As a result, the performance of WPT technology becomes very slow. Therefore, the WPT technology is improved and used "strong coupled" coils when the distance increases more than 2 m while charging wirelessly [2]. The two important WPT technologies are Inductive Power Transfer (IPT) and Capacitive Power Transfer (CPT). CPT is only applicable to low power applications with very short air gaps between 10–4 and 10–3 m, whereas IPT can be used for large air gaps around several meters, and its output power is much higher than CPT.

Figure 1 Compares between CPT and IPT, **Figure 1a** shows the power transfer capability versus gap distance with efficiency values. Furthermore the figure indicates both IPT and CPT can achieve ≥90% efficiency at kilowatt power levels in their respective gap ranges. **Figure 1b** plots the transmitter/receiver area versus throughput power with efficiency. The coupler area is the cross-sectional area through which magnetic or electric fields transfer energy. **Figure 1c** plots the output power density (output power divided by the gap volume) versus frequency [3]. WPT technologies can be applied in television, phone chargers, and induction heating, medical devices, pacemakers, radiofrequency identification, sensors, robotics, and deeply used in wireless charging for EV [4–11].

In CPT and IPT power transfers, the respective energy stored in a unitary volume of space is

$$W_e = \frac{1}{2}\varepsilon_0 E^2 \tag{1}$$

$$W_m = \frac{1}{2}\mu_0 H^2 \tag{2}$$

where E and H are the intensity of the electric and magnetic fields and, ε_0 and μ_0 are the permittivity and the permeability of the free space.

Health and safety, and economic impact are the critical points that should be considered in any technology of WPT for charging EVs or in any other applications. WPT technology for charge replacement, EVs can become gorgeous option. WPT charging has the advantage that it can make the charging process automated, suitable and safe for users and large scale introduction of WPT charging infrastructure

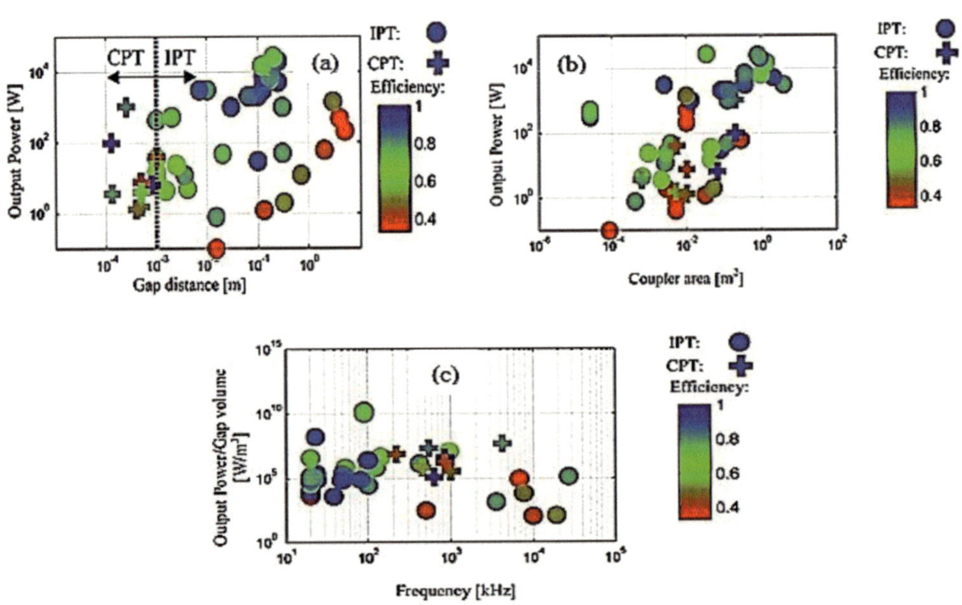

Figure 1.
Comparison between CPT and IPT technologies [3], (a) gap distance and output power with efficiency value, (b) coupler area and output power with efficiency value, (c) frequency and output power density with efficiency value.

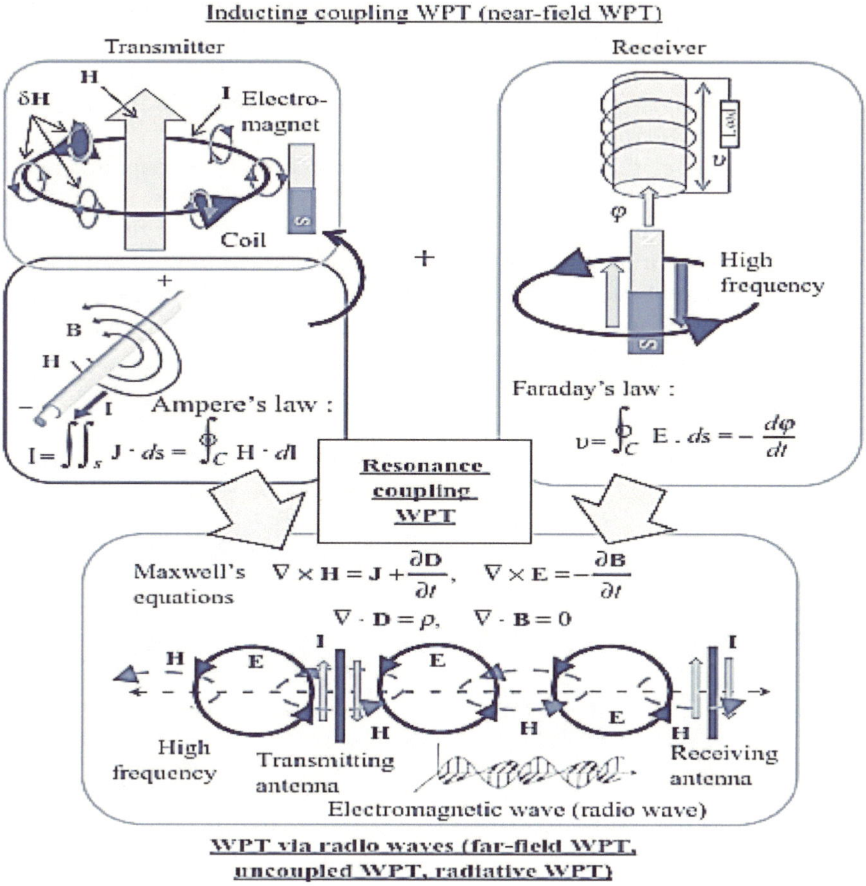

Figure 2.
Relationship between Ampere's law, Faraday's law, Maxwell's equations, inductive-coupling WPT, resonance-coupling WPT, and WPT via radio waves [13].

can help reduce the battery pack size and in turn make the EVs more proficient. However, all this cannot be knowledgeable by using traditional inductive chargers and WPT charging through large air gaps and least possible human interaction are required [12].

WPT systems are primarily classified as microwave, evanescent wave, magnetic resonance, electrical resonance, or electromagnetic induction methods. Scientists have newly proposed an electromagnetic induction method that is not premised on transformer coupling. It is revealed that the electric power transmission over a range including a magnetic field resonance method is enabled by adopting this method [13].

Ampere's law, Faraday's law, Maxwell's equations, the inductive-coupling WPT, the resonance-coupling WPT, and WPT via radio waves are the major theory that can describe the functionality of WPT technologies. **Figure 2** shows this relationship between these theories.

2. Basic theory of WPT system

When AC or DC electrical energy is transformed to high-frequency electrical energy by using a high-frequency inverter, the wireless feeding device (Tx.) releases electrical energy through a transmission device into space. Then, the receiving system (Rx.) converts the electrical power into DC in the recipient electrical apparatus. In addition, the efficiency of the electrical power transmission, medical and environmental influence of electromagnetic waves, and improvement the facile high speed charging, safe security, and energy storage density are essential limits should be considered when WPT is designed for EVs.

WPT systems are classified as microwave, evanescent wave, magnetic resonance, electrical resonance, or electromagnetic induction methods. **Figure 3** shows the relation between transmitted power and transmitted length of these methods for WPT systems. Additionally, WPT can be classified to the kind of Type 1,Type 2, Type 3 and Type 4 as shown in **Figure 4**. **Table 1** shows the relationship of Types 1–4 from the viewpoint of coupling mechanism, resonant mechanism,

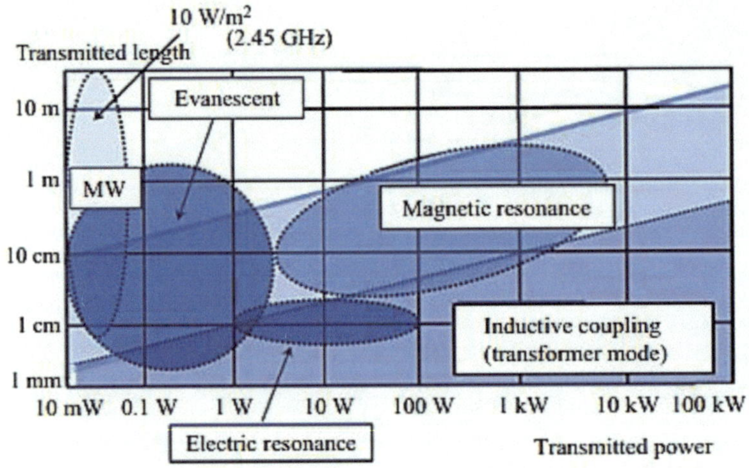

Figure 3.
Transmitted power and transmitted length of noncontact WPT methods [13].

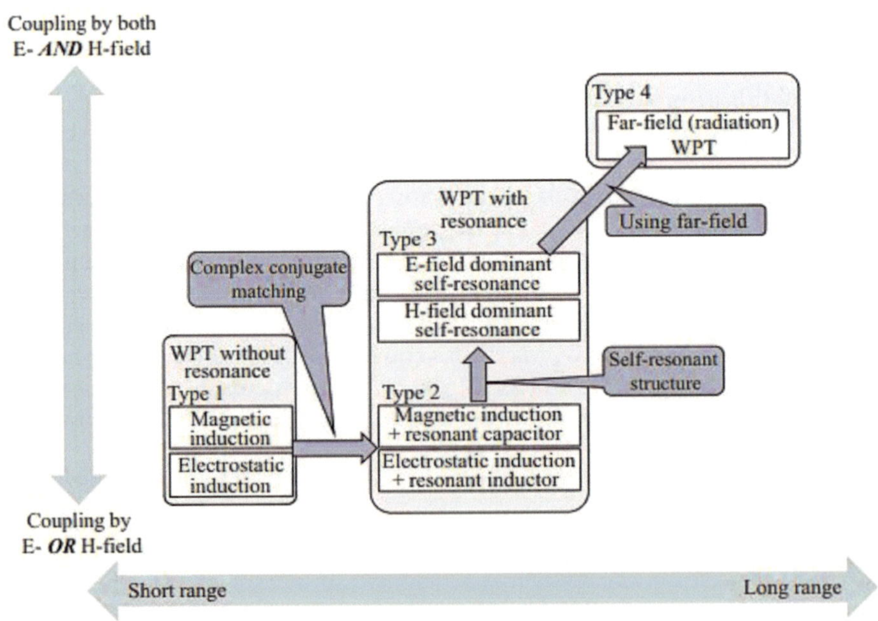

Figure 4.
Relationship of different kinds of WPT scheme.

Type		Coupling mechanism		Resonant mechanism	Impedance matching mechanism (feeding mechanism)	Schematic
		E-field	H-field			
1	Electrostatic induction	Yes	No	Power factor compensation may be considered as a resonant circuit	Not active following for load impedance	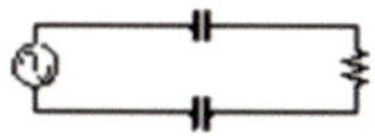
	Magnetic induction	No	Yes			
2	Coupled-resonant using electrostatic induction	Dominant	Negligible	Discrete reactance device is necessary for resonance	According to the load impedance or transmission distance, active following by circuit parameter in impedance matching circuit or transmission frequency is necessary to achieve simultaneous conjugate matching	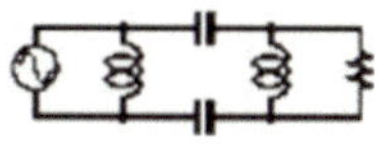
	Coupled-resonant using magnetic induction	Negligible	Dominant			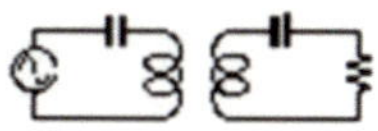
3	Coupled-resonant with self-resonant coupler (E-field dominant)	Dominant	Small, but not negligible	Coupler acts as resonator		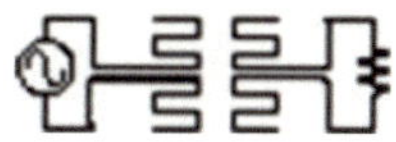
	Coupled-resonant with self-resonant coupler (H-field dominant)	Small, but not negligible	Dominant			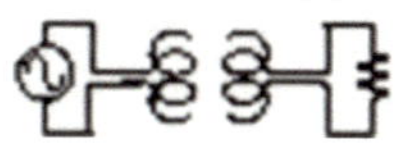
4	Far-field type (microwave WPT)	Coupling in far-field. Ratio of E-field to H-field is 377 Ω		Tx and Rx antennas resonate independently	Tx/Rx antennas are matched to the source/load independently	

Table 1.
Coupling mechanism, resonant mechanism, and impedance matching mechanism for various kinds of WPT.

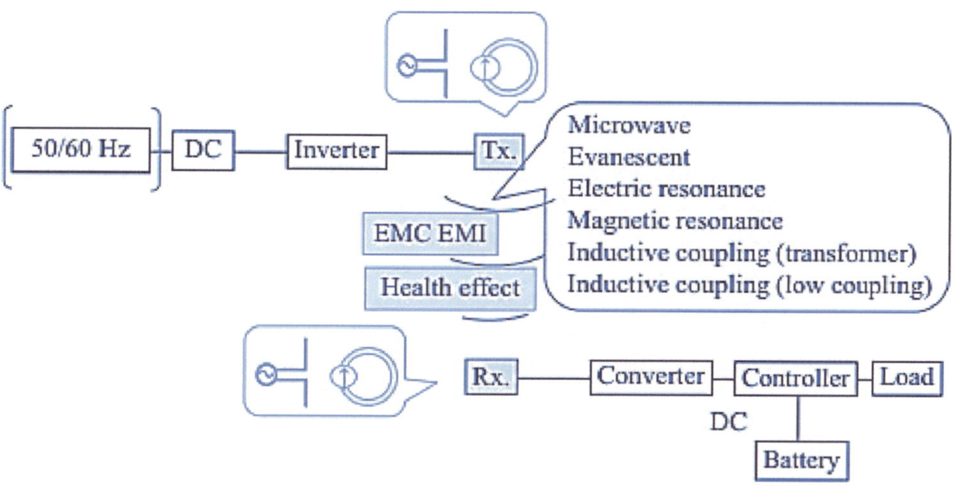

Figure 5.
Block diagram of WPT system.

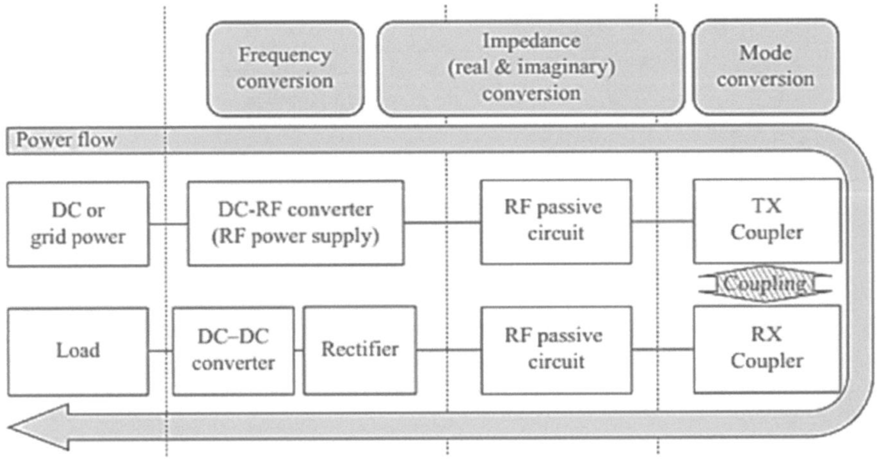

Figure 6.
Power flow in WPT system.

and feeding mechanism. **Figure 5** shows the general Block diagram of wireless power transfer system [13].

3. Power flow in WPT system

The block diagram of the Power flow in WPT system becomes as shown in **Figure 6**. RF inverter converts the frequency of the power. Typically, RF inverter also converts the voltage of the power. Considering that the power is P = VI and the impedance is Z = V/I, the RF inverter also converts the impedance of the power. The rectifier in Rx side also changes the frequency of the power.

4. WPT-systems for charging EVs

Figure 7 depicts the base components of a WPT system for EV charging. It consists of two prime sub-systems, one of which is existing underneath the road surface and the second is found into the vehicle underbody [14]. The first subsystem includes the source of energy, rectifier and high frequency inverter, primary compensation network and the primary/transmitter coil (Tx). The built subsystem in EVs, has the secondary/receiving coil (Rx) and secondary compensation network composes a resonance circuit that supplies into a high frequency rectifier, filter and the battery. The sub-systems are separated by an air gap. The distance between the two systems depends on the type of vehicle, ground clearance and road conditions such as pavement thickness. Usually the air gap is smaller than 0.4 m. Additionally, both sub-systems share information via a communication link.

The system given in **Figure 7** includes:

- Distribution network in the road has low-frequency AC power for various loads [15, 16].

- Rectifier and power factor correction (PFC).

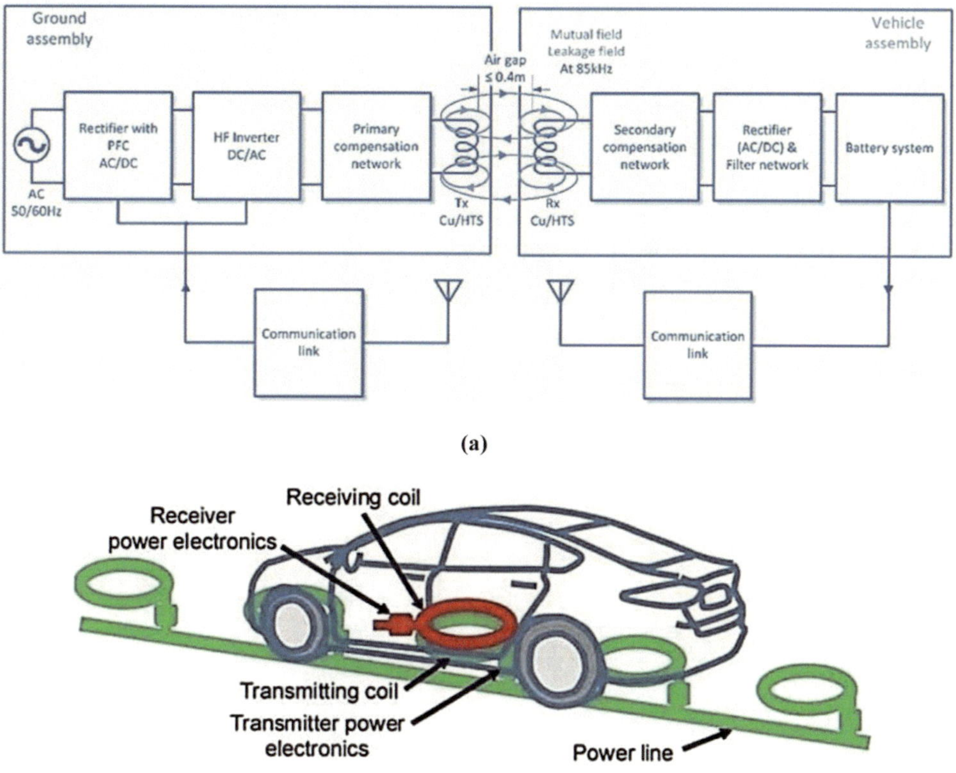

Figure 7.
*(**a**) The base components of a WPT system for (**b**) EV charging system.*

- High-frequency inverter converts the DC power to high frequency AC in the primary side. On the secondary side, the high-frequency AC energy is rectified to DC power and filtered to create a ripple free current that can charge the battery of EVs. The resonant frequency of the compensation topologies and coils determine the required switching frequency of the inverters. Usually used resonance frequencies for WPT EV-chargers are within a range of 20–100 kHz [17].

- Compensation networks are located between the high-frequency inverter and the primary coil in the ground assembly (GA), while between the secondary coil and the rectifier in the vehicle assembly (VA). **Table 2** summaries these networks and their efficiency. Power transfer efficiency with mutual inductance relationships for the basic compensation topologies SS, SP, PS, and PP are given in **Figure 8**. An optimal selection operation of the compensation topology based on the economics of the system is proposed in [18]. It concludes that SS and SP-compensation networks are the most appropriate topologies for high-power WPT systems. In addition, SS compensation needs less copper than the other compensation networks.

- Communication links. Evenly important to the power transfer system is the communication link between GA and VA. This also contains the communication to the GA grid connection to manage demand upon grid status.

Topology	Circuit	Power transfer efficiency at resonance $\eta = P_{out}/P_{in}$
SS	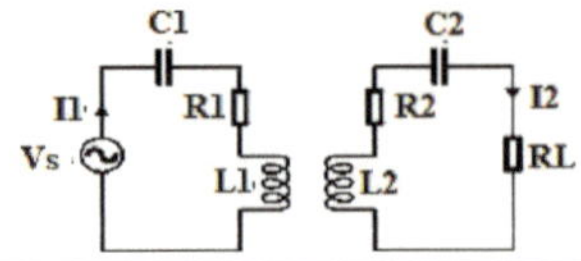	$\frac{R_L}{R_2+R_L+R_1\left(\frac{R_2+R_L}{\omega M}\right)^2}$
SP	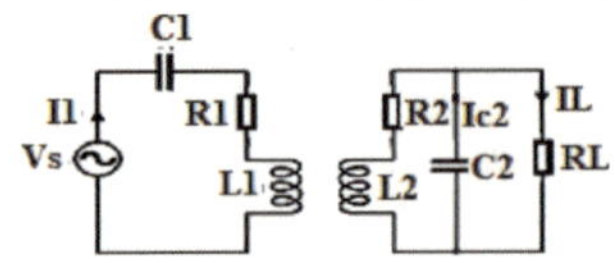	$\frac{R_L}{R_2+R_L+\frac{R_2 {R_L}^2}{(\omega L_2)^2}+\frac{R_1 {R_2}^2}{(\omega M)^2}+\frac{R_1\left(L_2\omega^2+\frac{R_L R_2}{\omega^2 L_2}\right)^2}{(\omega M)^2}}$
PS	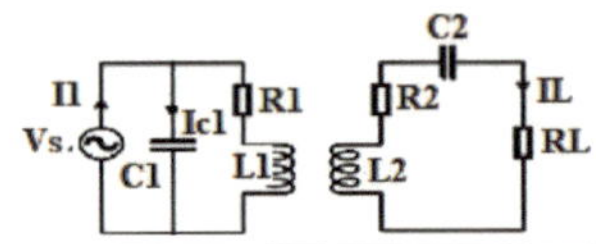	$\frac{R_L}{R_2+R_L+R_1\left(\frac{R_2+R_L}{\omega M}\right)^2}$
PP	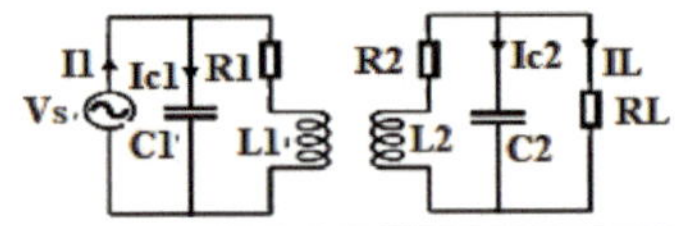	$\frac{R_L}{R_2+R_L+\frac{R_2 {R_L}^2}{(\omega L_2)^2}+\frac{R_1 {R_2}^2}{(\omega M)^2}+\frac{R_1\left(L_2\omega^2+\frac{R_L R_2}{\omega^2 L_2}\right)^2}{(\omega M)^2}}$
$\omega = \sqrt{\frac{1}{L_1 C_1}} = \sqrt{\frac{1}{L_2 C_2}} = \sqrt{\frac{1}{LC}}$ M: mutual inductance, ω: resonance frequancy		

Table 2.
SS, SP, PS, PP compensation topologies.

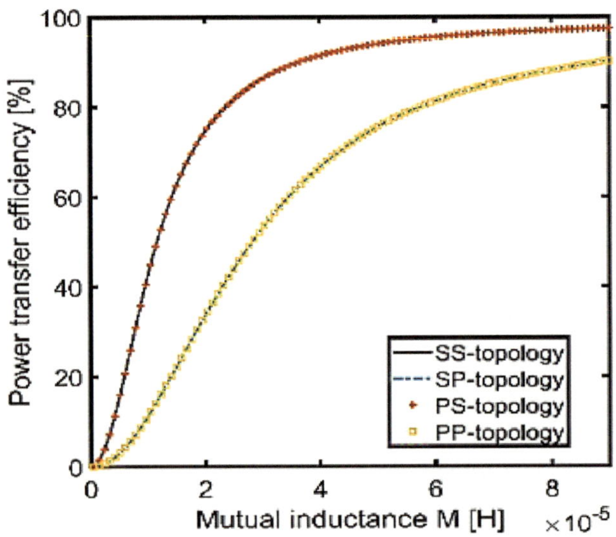

Figure 8.
Power transfer efficiency characteristics under varying mutual inductance [26].

- The core components of the WPT system are two coupled coils that allow power transfer via magnetic field [19]. Electric current flows through the primary coil and produces a time-varying magnetic field around it. In the nearness of the primary coil, the secondary coil intercepts the magnetic field, which induces a voltage. The value of induced voltage depends on the air gap length between these coils, the number of turns and the value of dφ/dt; φ is the magnetic flux.

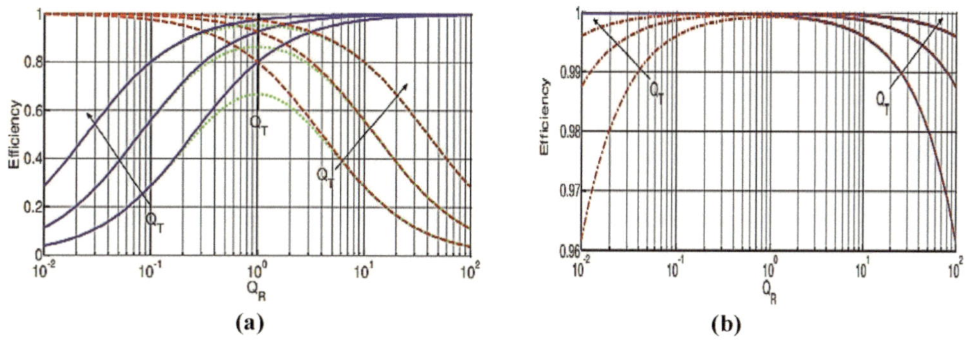

Figure 9.
Efficiency of resonant WPTSs with SS, SP, PS and PP topology [12].

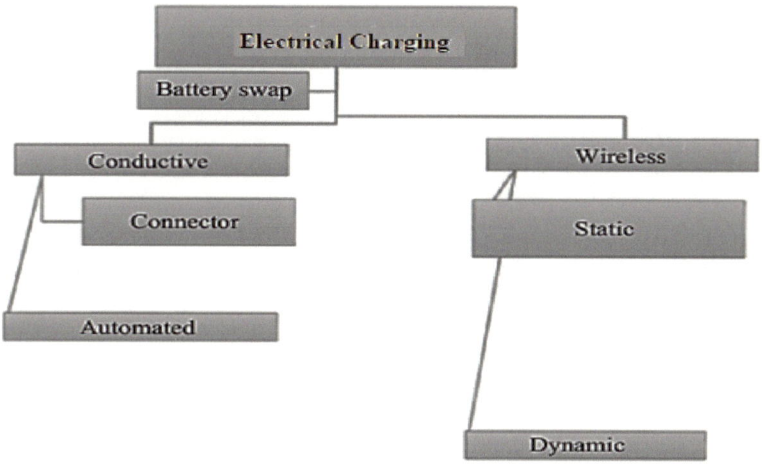

Figure 10.
Classification of charging method for EVs.

Moreover, different types of coil design are used in WPT, such that Circular, Flux pipe/flat solenoid, Bipolar, Tripolar, Zigzag, DD and DDQ [20–26].

Efficiency of resonant WPT with SS and SP topology, and of an inductive WPT vs. Q_R for different values of Q_T is reported in **Figure 9a**, while the efficiency of resonant WPTS with PS and PP topology is given in **Figure 9b**.

$$Q_{T,R} = \frac{\omega L_{1,2}}{R_{1,2}} \tag{3}$$

$$k = \frac{M}{\sqrt{L_1 L_2}} \tag{4}$$

$Q_{T,R}$: Transmitting and the receiving quality factor, respectively. k: Coupling coefficient. L1 and L2 are the s self-inductances of the transmitting and the receiving coils, M is their mutual inductance, and R1 and R2 are the coil resistances.

To charge or transfer energy into the vehicle electric storage device, many electrical charging methods are established and standardized. This methods are given if **Figure 10**.

Furthermore, there are many other many electrical charging methods uses different converters among different renewable sources [27, 28].

5. Major IEC/ISO standards for WPT for EV

1. **SAE J2954**: Wireless Power Transfer for Light-Duty Plug-In/Electric Vehicles and Alignment Methodology.

2. **SAE J2894/1**: Power Quality Requirements for Plug-In Electric Vehicle Chargers.

3. **SAE J2847/6**: Communication between Wireless Charged Vehicles and Wireless EV Chargers.

4. **SAE J2931/6**: Signaling Communication for Wirelessly Charged Electric Vehicles.

5. **ICNIRP 2010**: ICNIRP Guidelines for limiting exposure to time-varying electric and magnetic fields (1 Hz – 100 kHz).

6. **ISO 14117:2012:** Active implantable medical devices – Electromagnetic compatibility – EMC test protocols for implantable cardiac pacemakers, implantable cardioverter defibrillators and cardiac resynchronization devices.

7. **ISO/PAS 19363:2017**: Electrically propelled road vehicles – Magnetic field wireless power transfer – Safety and interoperability requirements.

8. **ISO 15118:** Road vehicles – Vehicle to grid communication interface.

9. **IEC61980–1:** Electric vehicle wireless power transfer (WPT) systems – Part 1: General requirements.

6. Conclusion

In this chapter, several WPT technologies are presented and explained the importance of WPT. In recent years, WPT technology gains enormous attention due to its advantages. WPT systems are classified as microwave, evanescent wave, magnetic resonance, electrical resonance, or electromagnetic induction methods. Moreover, the prospective EV wireless charging applications are also highlighted, and different coupling techniques are discussed in this chapter. Inductive coupling, magnetic resonance coupling, and microwave are the main EV wireless charging techniques. The chapter also provided detailed information on the manufacturing and configuring the magnetic resonance coupling based wireless charging system. This chapter deals with the basic overview of the present and future scenarios of EVs. After discussing the scenarios, various concerns regarding battery types and charging methodologies are discussed as well.

Author details

Nassim Iqteit[1]*, Khalid Yahya[2] and Sajjad Ahmad Khan[3]

1 Department of Electrical Engineering, Palestine Polytechnic University, Palestine

2 Department of Mechatronic Engineering, Istanbul Gelisim University, Turkey

3 Department of Electronics and Communication Engineering, Istanbul Technical University (ITU), Turkey

*Address all correspondence to: nassim_eng83@ppu.edu

References

[1] Tesla N. Apparatus for Transmitting Electrical Energy. New York, USA Patent 1119732; 1914.

[2] Kurs A, Karalis A, Moffatt R, Joannopoulos J, Fisher P, Soljacic M. Wireless Power Transfer via Strongly Coupled Magnetic Resonances. Science 2007;317(5834):83–6.

[3] Dai J, Ludois DC. A Survey of Wireless Power Transfer and a Critical Comparison of Inductive and Capacitive Coupling for Small Gap Application. IEEE Trans Power Electron 2015;30(11): 6017–29.

[4] Kim J, Son H-c, Kim D-h, Park Y-j. Optimal Design of a Wireless Power Transfer System with Multiple Self-Resonators for an LED TV. IEEE Trans Consum Electron 2012;58(3):775–80.

[5] Moon J, Hwang H, Jo B, Kwon C-K, Kim T-G, Kim S-W. Design and Implementation of a high-efficiency 6.78MHz resonant wireless power transfer system with a 5W fully integrated power receiver. IET Power Electron 2017;10(5):577–8.

[6] Yengi Y., Khan S. A., and Küçük K., Design and performance analysis of information centric network for Internet of Things, 25th Signal Processing and Communications Applications Conference (SIU), Antalya, 2017, pp. 1–4.

[7] Agrawal K, Jegadeesan R, Guo Y-X, Thakor NV. Wireless Power Transfer Strategies for Implantable Bioelectronics: methodological Review. IEEE Rev Biomed Eng 2017;10:1–28.

[8] Erdoğan H., Küçük K., Khan S. A., Endüstriyel IoT Bulut Uygulamaları için Düşük Maliyetli Modbus/MQTT Ağ Geçidi Tasarımı ve Gerçekleştirilmesi . Bilecik Şeyh Edebali Üniversitesi Fen Bilimleri Dergisi, 2020, (7) (1), 170–183.

[9] Trevisan R, Costanzo A. A UHF Near-Field Link for Passive Sensing in Industrial Wireless Power Transfer Systems. IEEE Trans Microw Theory Tech 2016;64(5):1634–43.

[10] Sugino M, Masamura T. The wireless power transfer systems using the Class E push-pull inverter for industrial robots. In: IEEE Wireless Power Transfer Conference (WPTC), Taipei, Taiwan; 2017.

[11] Li S, Mi CC. Wireless power transfer for electric vehicle applications. Journal Emerg Sel Top Power Electron 2015;3 (1):4–17.

[12] Naik Mude K. Wireless Power Transfer for Electric Vehicle. [Phd thesis], University of Padova, Italy.

[13] Shinohara N. Wireless Power Transfer. Stevenage: IET; 2018.

[14] Society of Automotive Engineers (SAE). J2954 - Wireless Power Transfer for LightDuty Plug-In/ Electric Vehicles and Alignment Methodology, SAE international, Warrendale, USA; 2016.

[15] Nassim I, Basa Arsoy A, & Çakır B. (2018). Load Profile-Based Power Loss Estimation for Distribution Networks. Electrica, 18(2), 275–283.

[16] Nassim I, Basa A, Çakir B. A Simple Method to Estimate Power Losses in Distribution Networks. In Proceedings of the 10th International Conference on Electrical and Electronics Engineering (ELECO),Bursa, Turkey, 30 November–2 December 2017.

[17] Vilathgamuwa D, Sampath J. Wireless Power Transfer (WPT) for Electric Vehicles (EVs) - Present and Future Trends. In: Rajakaruna S, Gosh A, Shahnia F, editors. Plug In Electric Vehicles in Smart Grids -

Integration. Techniques Singapore: Springer Science+Business Media Singapore; 2015. p. 33–61.

[18] Sallán J, Villa JL, Llombart A, Sanz JF. Optimal Design of ICPT Systems Applied to Electric Vehicle Battery Charge. IEEE Trans Ind Electron 2009;56(6):2140–9.

[19] Nassim I, and K Yahya. Simulink model of transformer differential protection using phase angle difference based algorithm", Inter. J. of Power Electronics and Drive Systems, vol. 11, no. 2, pp. 1088–1098, 2020.

[20] Budhia M, Covic GA, Boys JT. Design and Optimization of Circular Magnetic Strucutres for Lumped Inductive Power Transfer Systems. IEEE Trans Power Electron 2011;26(11): 3096–108.

[21] Nagatsuka Y, Ehara N, Kaneko Y, Abe S, Yasuda T. Compact Contactless Power Transfer System for Electric Vehicles. In: International Power Electronics Conference - ECCE ASIA -, Hokkaido, Japan; 2010.

[22] Zaheer A, Kacprzak D, Covic GA. A Bipolar Receiver Pad in a Lumped IPT System for Electric Vehicle Charging Applications. In: IEEE Energy Conversion Congress and Exposition (ECCE), Raleigh, USA; 2012.

[23] Kim S, Covic GA, Boys JT. Comparison of Tripolar and Circular Pads for IPT Charging Systems. IEEE Trans Power Electron 2018;33(7): 6093–103.

[24] Alam MM, Mekhilef S, Seyedmahmoudian M, Horan B. Dynamic Charging of Electric Vehicle with Negligible Power Transfer Fluctuation. Energies 2017;10(701).

[25] Budhia M, Boys JT, Covic GA, Huang C-Y. Development of a Single-Sided Flux Magnetic Coupler for Electric Vehicle IPT Charging Systems. IEEE Trans Ind Electron 2013;60(1): 318–28.

[26] Machura P, Li Q. A critical review on wireless charging for electric vehicles. Renewable and Sustainable Energy Reviews. 2019;104:209–234. doi: 10.1016/j.rser.2019.01.027

[27] K. Yahya, M. Z. Bilgin, T. Erfidan, and B. Cakir, “Improving the performance of the MPPT for thermoelectric generator system by using Kalman filter,” in Proc. 5th Int. Conf. Electr. Electron. Eng. (ICEEE), Istanbul, Turkey, May 2018, pp. 129–132, doi: 10.1109/ICEEE2.2018.8391315.

[28] Yahya K, Bilgin MZ, Erfidan T. Practical implementation of maximum powertracking based short-current pulse method for thermoelectric generators systems. JPower Electron 2018;18(4):1201–10.

Chapter 4

Theoretical and Practical Design Approach of Wireless Power Systems

Vladimir Kindl, Michal Frivaldsky, Jakub Skorvaga and Martin Zavrel

Abstract

The paper introduces the main issues concerned with the conceptual design process of wireless power systems. It analyses the electromagnetic design of the inductive magnetic coupler and proposes the key formulas to optimize its electrical parameters for a particular load. For this purpose, a very detailed analysis is given focusing on the mathematical concept procedure for determination of the key factors influencing proper coupling coils design. It also suggests basic topologies for conceptual design of power electronics and discusses its proper connection to the grid. The proposed design strategy is verified by experimental laboratory measurement including analyses of leakage magnetic field.

Keywords: wireless power transfer, coil design, analytical approach, electromagnetic field, efficiency, optimization, shielding

1. Introduction

There are currently many methods, links and approaches for wireless power transmission. Each of the available solutions is characterized by its advantages and disadvantages, which result in their application [1–3].

Inductive coupling is currently the most widely used method of wireless energy transfer. This method works on the principle of an air transformer with a tight magnetic coupling of the primary and secondary windings. The energy exchange between two or more coils takes place by means of an inductive current Φ, i.e. by means of an induced voltage. The main disadvantage of inductive coupling is the transmission distance, which ranges from millimeters to several centimeters [4].

Resonance compensation is a specific case of inductive coupling. Resonant compensation is used in cases where it is necessary to achieve impedance dependence on frequency. Resonant compensation is provided by adding a capacitive member to the primary as well as the secondary coil. After applying a magnetic field with a suitably selected frequency, the phenomenon of mutual interference of the impedance of the coil and the capacitor occurs, which ideally ensures zero phase shift against the current flowing through the primary coil. For resonant compensation, there are four configurations of the primary and secondary side of the wireless charging system [5, 6].

The system using resonant coupling fully compensates for the scattering fields of the coupling coils, thus significantly extending the working distance while maintaining high energy transfer efficiency. Thanks to its advantages, resonant coupling is used mainly in the field of electromobility, where it allows charging with high power and in the case of a variable load, it can be easily frequency-adjusted for optimal efficiency [7, 8].

Energy transmission through capacitive coupling is currently used relatively little due to limitations on the transmission distance, which is limited by the level of tenths of a millimeter. This method is mainly used for charging consumer electronics such as tablets, laptops and more. They also have great potential in the field of medicine for charging various implants. Capacitive coupling is a phenomenon occurring between all conductive objects, i.e. between systems between which there is a mutual difference of potentials and between them there is an environment with a positive dielectric constant (permittivity) [9–11].

This work aims to point out the main design issues related to wireless power transmission and demonstrate their operational characteristics. An important aspect in this area is undoubtedly interaction of living organisms with a strong electromagnetic field, and therefore it is necessary to pay attention also legislation and hygiene standards [12–15]. Another goal is to provide a clear mathematical description of the system using intuitive methods for circuit analysis. Mathematical models must consider, in addition to the coupling itself, the inverters (inverter and rectifier) on the primary and secondary side of the system. An equally important goal of the work is experimental verification of all achieved theoretical conclusions. For this purpose, it was necessary to develop a prototype of a WPT charging system capable to supply sufficient power needed to charge conventional electric car. The text is supplemented by accompanying graphics that illustrates efficiency characteristics and also analysis of the spatial distribution of the electromagnetic field at different states of the system.

2. Basic resonant coupling techniques

There are four basic configurations of the primary and secondary side of the WPT system to realize resonant compensation of the leakage inductance [16–18]. This chapter focuses on investigating the properties of possible compensation methods, which include serial-serial, serial-parallel, parallel-serial, parallel–parallel.

Parameter	Value	
R_1	0.45 Ω	
R_2	0.45 Ω	
L_1	145.6 μH	
L_2	145.6 μH	
C_1	3.1 nF	
C_2	3.1 nF	
k	0÷0.1	
R_Z	Series compensation: 5÷200 Ω	Parallel comp.: 5 Ω÷2.5 kΩ
U_b	100 V	

Table 1.
Circuit parameters for the evaluation of compensation techniques.

The analysis of individual configurations is further provided, while and examples of characteristics derivations are based on the circuit parameters listed in **Table 1**.

2.1 Series-series compensation

Serial-series compensation uses two external capacitors C_1 and C_2 connected in series with the primary and secondary windings. The circuit model of the system is shown in **Figure 1**.

The system is powered by an inverter with a rectangular voltage profile with amplitude um1, and therefore the circuit must be described by a system of integrodifferential equations forming a full-fledged dynamic model.

$$\begin{aligned} -u_1 + \frac{1}{C_1}\int_0^t i_1 dt + u_{C_1(0)} + R_1 i_1 + L_1\frac{di_1}{dt} + M\frac{di_2}{dt} = 0 \\ -L_2\frac{di_2}{dt} - M\frac{di_1}{dt} - R_2 i_2 - \frac{1}{C_2}\int_0^t i_2 dt - u_{C_2(0)} - R_Z i_2 = 0 \end{aligned} \tag{1}$$

All models will be derived for the fundamental harmonic and therefore we can use Eq. (1) to describe the model, while the inverter voltage can be considered in the form of (2).

$$\acute{U}_1 = \frac{2\sqrt{2}}{\pi} u_{m1} \tag{2}$$

The solution we get loop currents, from which it is possible to further determine all operating variables of the system (3).

$$\begin{bmatrix} \acute{I}_{S1} \\ \acute{I}_{S2} \end{bmatrix} = \begin{bmatrix} R_1 + j\left(\omega L_1 - \dfrac{1}{\omega C_1}\right) & -j\omega M \\ -j\omega M & R_2 + R_Z + j\left(\omega L_2 - \dfrac{1}{\omega C_2}\right) \end{bmatrix} \left[\begin{bmatrix} \acute{U}_1 \\ 0 \end{bmatrix}\right] \tag{3}$$

For a better idea, we draw the efficiency and power on the load depending on the frequency and the coupling factor, respectively the load. This creates two pairs of maps in which two functions are plotted separately:

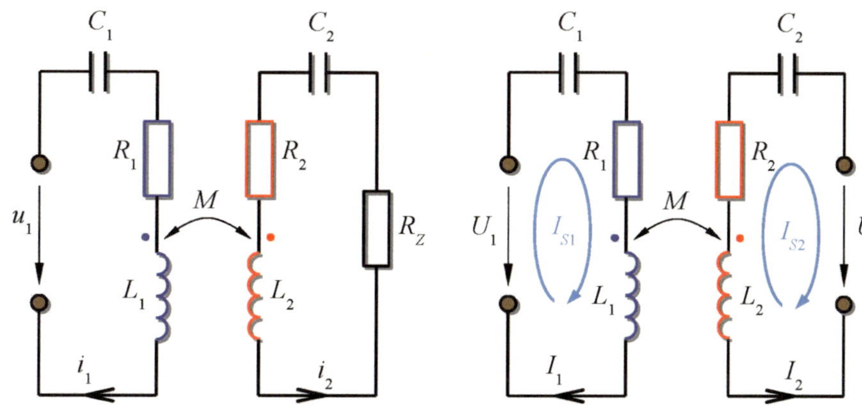

Figure 1.
Simplified equivalent circuit for series–series compensation, left – Circuit with initial variables, right – Circuit suited for loop current analysis.

$$P_2 = f(f_{sw}, k), P_2 = f(f_{sw}, R_Z), \text{and } \eta = f(f_{sw}, k), \eta = f(f_{sw}, R_Z) \tag{4}$$

The first map will consider a constant load, which will be set as optimal for the working distance corresponding to $k = 0.1$ [19]. The optimal load for the map was determined based on the relationship:

$$R_{Zs-efficiency} = \sqrt{\frac{R_2(R_1R_2 + M^2\omega^2)}{R_1}} \doteq 22\Omega. \tag{5}$$

The resulting maps are shown in **Figure 2**.

2.2 Series-parallel compensation

Serial-parallel compensation uses two resonant capacitors connected in series with a coupling coil on the primary side and in parallel on the secondary side (**Figure 3**). In the left part you can see the diagram for the dynamic model and in the right part its simplification for the supply of the harmonic course of the voltage.

Unlike the previous circuit, it is now necessary to compile three equations of three unknowns. The integrodifferential form is given in the (6).

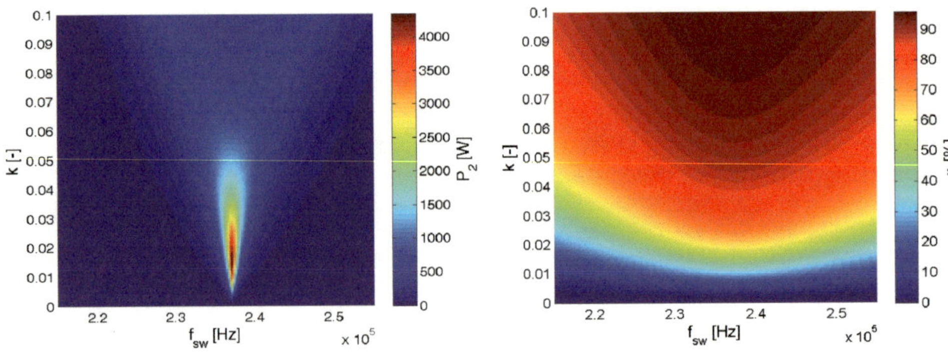

Figure 2.
Dependency of output power (left) and system efficiency (right) on frequency and power transfer distance for S-S compensation.

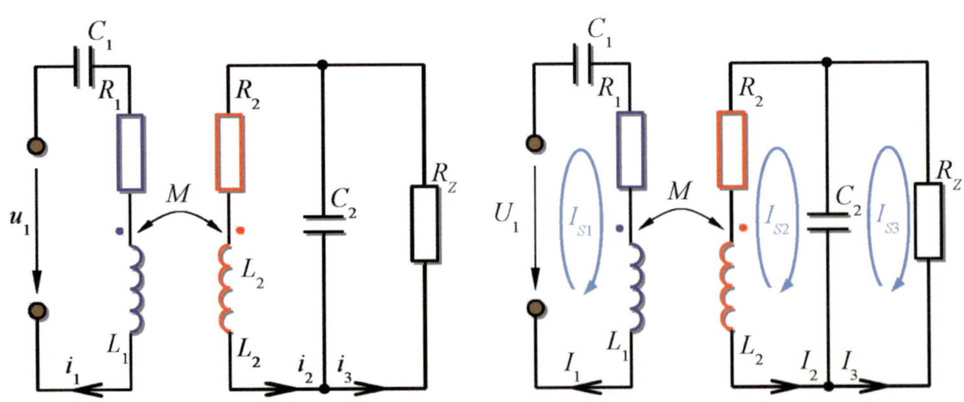

Figure 3.
Simplified equivalent circuit for series–parallel compensation, left – Circuit with initial variables, right – circuit suited for loop current analysis.

$$-u_1 + \frac{1}{C_1}\int_0^t i_1 dt + u_{C_1(0)} + R_1 i_1 + L_1\frac{di_1}{dt} + M\frac{di_2}{dt} = 0$$
$$-L_2\frac{di_2}{dt} - M\frac{di_1}{dt} - R_2 i_2 + \frac{1}{C_2}\int_0^t (i_3 - i_2)dt + u_{C_2(0)} = 0 \quad (6)$$
$$-\frac{1}{C_2}\int_0^t (i_3 - i_2)dt - u_{C_2(0)} - R_Z i_3 = 0$$

For the calculation by the loop current method, the equations take the form (7).

$$\begin{bmatrix} \acute{I}_{S1} \\ \acute{I}_{S2} \\ \acute{I}_{S3} \end{bmatrix} = \begin{bmatrix} R_1 + j\left(\omega L_1 - \frac{1}{\omega C_1}\right) & -j\omega M & 0 \\ -j\omega M & R_2 + j\left(\omega L_2 - \frac{1}{\omega C_2}\right) & \frac{j}{\omega C_2} \\ 0 & \frac{j}{\omega C_2} & \frac{-j}{\omega C_2} + R_Z \end{bmatrix} \begin{bmatrix} \acute{U}_1 \\ 0 \\ 0 \end{bmatrix} \quad (7)$$

The optimal load is then given by the Eq. (8)

$$R_{Zp-efficiency} = \frac{L_2\omega\sqrt{R_1R_2^2 + \omega^2\left(L_2^2R_1 + M^2R_2\right)}}{\sqrt{R_2\left(R_1R_2 + M^2\omega^2\right)}} \doteq 2176\Omega. \quad (8)$$

The resulting waveforms are graphically summarized in **Figure 4**.

2.3 Parallel-series compensation

Parallel–series compensation is practically only like the previous variant, which uses two resonant capacitors connected in parallel with the coupling coil on the primary side and in series on the secondary side. The circuit model is apparent from **Figure 5**.

As in the previous case, it is enough to compile three equations of three unknowns for the description (9).

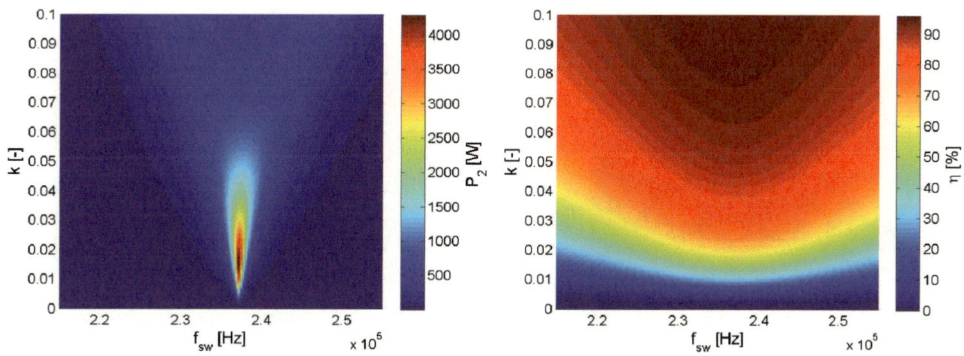

Figure 4.
Dependency of output power (left) and system efficiency (right) on frequency and power transfer distance for S-P compensation.

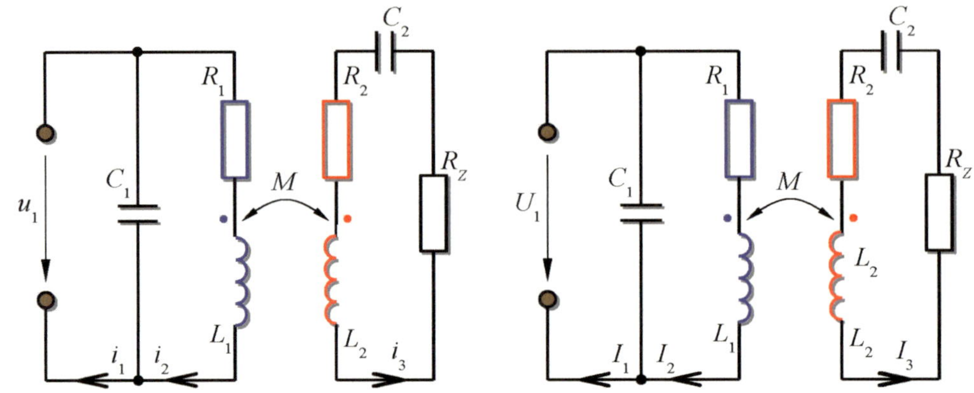

Figure 5.
Simplified equivalent circuit for parallel - series compensation, left – Circuit with initial variables, right – Circuit suited for loop current analysis.

$$\begin{aligned}
&-u_1+\frac{1}{C_1}\int_0^t (i_1-i_2)dt+u_{C_1(0)}=0\\
&-\frac{1}{C_1}\int_0^t (i_1-i_2)dt-u_{C_1(0)}+R_1 i_2+L_1\frac{di_2}{dt}+M\frac{di_3}{dt}=0\\
&-L_2\frac{di_3}{dt}-M\frac{di_2}{dt}-R_2 i_3-R_Z i_3=0
\end{aligned} \tag{9}$$

The equations below for the calculation by the loop method can be expressed as (10).

$$\begin{bmatrix} \acute{I}_{S1}\\ \acute{I}_{S2}\\ \acute{I}_{S3}\end{bmatrix}=\begin{bmatrix} \frac{-j}{\omega C_1} & \frac{j}{\omega C_1} & 0\\ \frac{j}{\omega C_1} & R_1+j\left(\omega L_1-\frac{1}{\omega C_1}\right) & -j\omega M\\ 0 & -j\omega M & R_2+R_Z+j\left(\omega L_2-\frac{1}{\omega C_2}\right)\end{bmatrix}\begin{bmatrix}\acute{U}_1\\ 0\\ 0\end{bmatrix} \tag{10}$$

From the above equations we again obtain the courses of all-important operating variables, shown in **Figure 5**. As in the previous case, the system achieves maximum efficiency at the optimized load (Rz = 22 Ω), while the transmitted power is significantly lower compared to the achievable value (**Figure 6**).

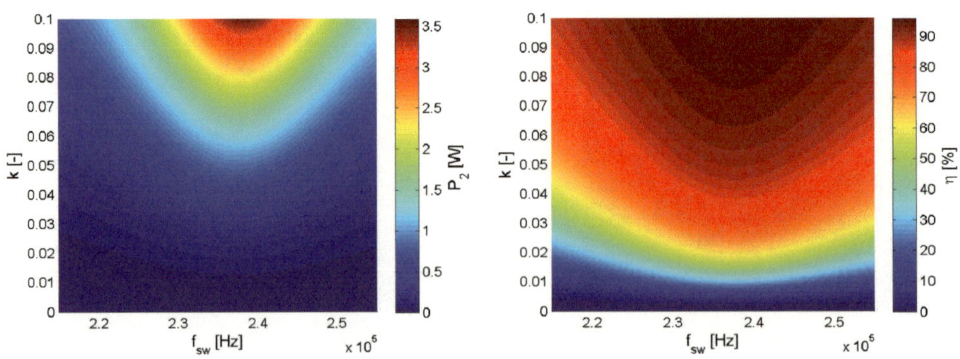

Figure 6.
Dependency of output power (left) and system efficiency (right) on frequency and power transfer distance for P-S compensation.

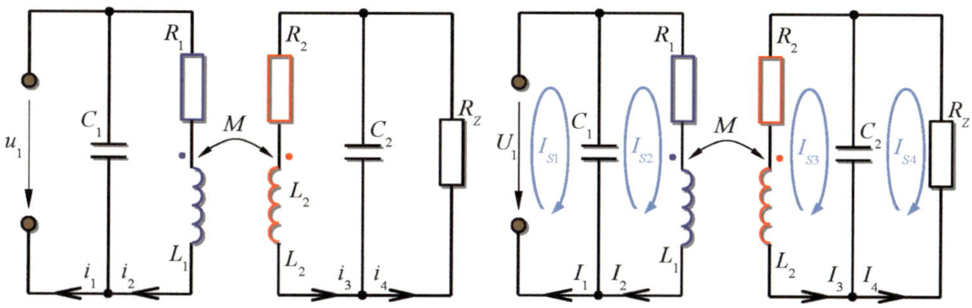

Figure 7.
Simplified equivalent circuit for parallel - parallel compensation, left – Circuit with initial variables, right – Circuit suited for loop current analysis.

2.4 Parallel-parallel compensation

Parallel–parallel compensation uses two resonant capacitors connected in parallel to both coupling coils. The circuit model can be seen in **Figure 7**.

Unlike previous models, in this case it is necessary to compile four equations with four unknowns, the integrodifferential form is represented by the Eqs. (11).

$$\begin{aligned}
&-u_1 + \frac{1}{C_1}\int_0^t (i_1 - i_2)dt + u_{C_1(0)} = 0 \\
&-\frac{1}{C_1}\int_0^t (i_2 - i_1)dt - u_{C_1(0)} + R_1 i_2 + L_1\frac{di_2}{dt} + M\frac{di_3}{dt} = 0 \\
&-L_2\frac{di_3}{dt} - M\frac{di_2}{dt} - R_2 i_3 + \frac{1}{C_2}\int_0^t (i_4 - i_3)dt + u_{C_2(0)} = 0 \\
&-\frac{1}{C_2}\int_0^t (i_4 - i_3)dt - u_{C_2(0)} - R_Z i_4 = 0
\end{aligned} \tag{11}$$

After stabilization we can rewrite the equations into the form (12).

$$\begin{bmatrix} \acute{I}_{S1} \\ \acute{I}_{S2} \\ \acute{I}_{S3} \\ \acute{I}_{S4} \end{bmatrix} = \begin{bmatrix} \frac{-j}{\omega C_1} & j\frac{1}{\omega C_1} & 0 & 0 \\ \frac{j}{\omega C_1} & R_1 + j\left(\omega L_1 - \frac{1}{\omega C_1}\right) & -j\omega M & 0 \\ 0 & -j\omega M & R_2 + j\left(\omega L_2 - \frac{1}{\omega C_2}\right) & \frac{j}{\omega C_2} \\ 0 & 0 & \frac{j}{\omega C_2} & R_Z - \frac{j}{\omega C_2} \end{bmatrix} \begin{bmatrix} \acute{U}_1 \\ 0 \\ 0 \\ 0 \end{bmatrix} \tag{12}$$

Graphical interpretations of the characteristics are shown in **Figure 8**.

2.5 Overall comparison

Based on the above results, we can compile a table that compares the key properties of individual compensation methods (**Table 2**). In the evaluation, we consider a system operating to the optimal load at a distance corresponding to the coupling factor $k < 0.1$. The supply voltage is the same for all compensation topologies and is not regulated.

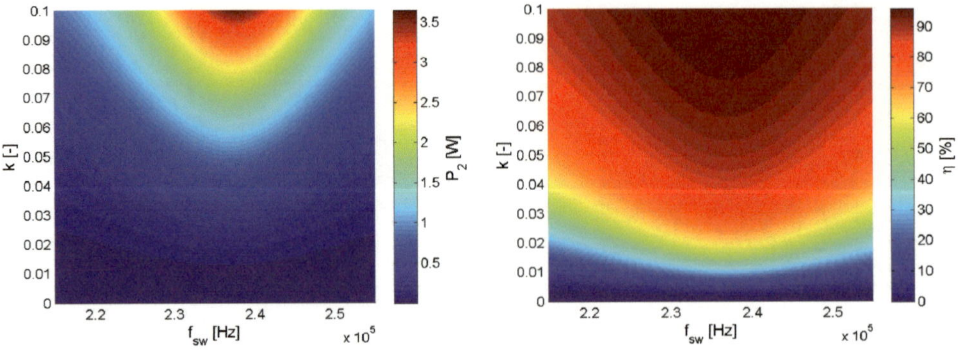

Figure 8.
Dependency of output power (left) and system efficiency (right) on frequency and power transfer distance for P–P compensation.

Criteria	*Compensating topology*			
	S-S	*S-P*	*P-S*	*P–P*
Source output type	current	voltage	voltage	Current
Power transfer ability	higher	higher	lower	lower
Max power and efficiency overlap	No	No	Partial	Partial
Optimum load value	lower	higher	lower	higher
Frequency sensitivity	higher	higher	lower	lower

Table 2.
Comparisons of key attributes of individual compensation techniques.

Serial-to-series compensation acts as a current source over the monitored operating distance range, allowing higher power to be delivered to the load compared to other topologies. The disadvantage is the minimum overlap of work areas with maximum system performance and efficiency.

Serial-parallel compensation in this case does not bring any significant operational benefits, the only difference lies in the higher values of the optimal load. Unlike the previous solution, the circuit acts as a voltage source.

Parallel–series compensation offers partial overlap of work areas with maximum performance and efficiency. However, the theoretically achievable transmitted power values are significantly lower compared to the two previous configurations. The circuit has a voltage output and is much less sensitive to frequency detuning.

Parallel–parallel compensation offers current output with better coverage of areas of maximum power and efficiency along with better frequency stability. However, the transmitted powers are very low, as with parallel–series compensation.

3. Identification of key system parameters and analytical approach for design of coupling elements

In the previous chapter, the principal characteristics regarding basic modifications of the main circuits for wireless power systems were derived and described. All models, in some form, use concentrated parameters of spare electrical circuits. These parameters can be determined basically in two ways, i.e. by calculation and by measurement, while the measurement can be used only if the analyzed system already physically exists. On the other side the calculations offer the possibility of optimization within design procedure.

Within next text, the systematic analytical procedure for calculation of key parameters of coupling elements for wireless power transfer is described, whereby the application area for any compensation technique can be considered here.

3.1 Mutual inductance of planar coils

In practice, circular or spiral coils are most often used for high-frequency purposes. The reason is the high gradients of the electric field, which arise on all structural edges of the coil in the case of parallel resonance. These gradients significantly worsen the quality factor and thus the operating characteristics of the resulting system [20, 21].

The mutual inductance of different clusters of air coils of spiral shape can be based on the application of the analytical rule for the magnetic vector potential in cylindrical coordinates (13).

$$A_\varphi[r,\varphi,z] = \frac{\mu_0 I}{4\pi} r_1 \int_0^\pi \frac{\cos(\varphi')}{\sqrt{r^2 + r_1^2 - 2rr_1\cos(\varphi') + (z - z')^2}} d\varphi', \tag{13}$$

The resulting relationship is based on the direct application of Biot-Savart's law. In technical practice, these integrals are abundant, and therefore considerable attention has been paid to their enumeration in the past. The literature defines three basic types of these (elliptic) integrals, which can be combined with each other and easily converted to any special case.

The Eq. (14) was determined based on Eq. (13).

$$A_\varphi[r,\varphi,z] = \frac{\mu_0 I}{4\pi} r_1 \frac{2}{\sqrt{r_1 r}} \left[\left(\frac{2}{k_I} - k_I\right) K(k_I) - \frac{2}{k_I} E(k_I)\right] \tag{14}$$

Where $K(k_I)$ a $E(k_I)$ are elliptical integrals of first and second type and have following forms:

$$K(k_I) = \int_0^{\frac{\pi}{2}} \frac{d\varphi}{\sqrt{1 - k_I^2 \sin^2(\varphi)}} a E(k_I) = \int_0^{\frac{\pi}{2}} \left[\sqrt{1 - k_I^2 \sin^2(\varphi)}\right] d\varphi. \tag{15}$$

Module of these integrals is determined using (16).

$$k_I = \sqrt{\frac{4rr_1}{(r + r_1)^2 + (z - z')^2}} \tag{16}$$

The derived relations correspond to a simplified geometry, where only a coaxial arrangement is considered [22]. In order to be able to calculate the mutual inductance of the coils of general geometry and arrangement, we must introduce the possibility of deflection, see **Figure 9** left.

For this special case, the procedure for modifying the previous equations was indicated in [22, 23]. For the mutual inductance of the two turns from **Figure 9** (left) we can write according to **Figure 9** (right) next Eq. (17).

$$M_{ij} = \frac{2\mu_0}{\pi} \sqrt{r_{1i} r_{2j}} \frac{\int_0^\pi \left[\cos() - \frac{d}{r_{2y}} \cos(\varphi)\right] \left[\left(1 - \frac{k_{II}^2}{2}\right) K(k_{II}) - E(k_{II})\right] d\varphi}{k_{II}\sqrt{3}}, \tag{17}$$

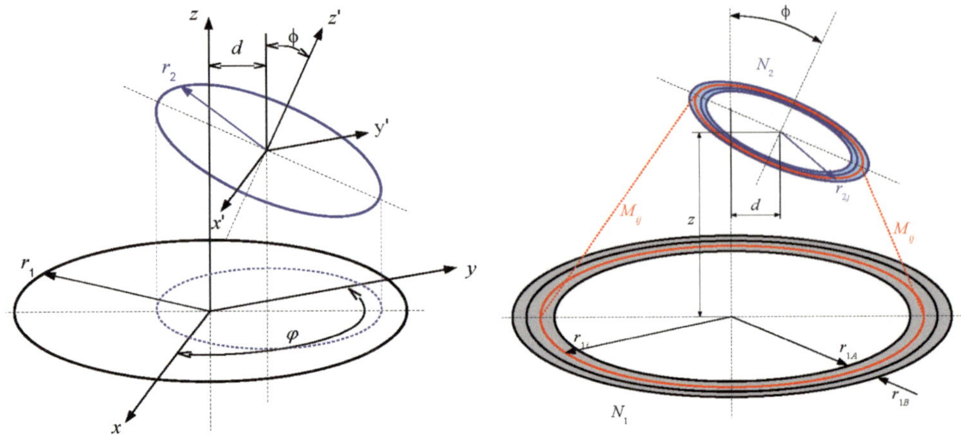

Figure 9.
Displacement of coil's turns and presentation of the situation for determination of mutual inductance for circular coils.

where

$$\sqrt{1 - sin^2() \, cos^2(\varphi) + \left(\frac{d}{r_{2j}}\right)^2 - \frac{2d}{r_{2j}} cos() \, cos(\varphi)}, \tag{18}$$

and then

$$k_{II} = \sqrt{\frac{4\frac{r_{2j}}{r_{1i}}}{\left(1 + \frac{r_{2j}}{r_{1i}}\right)^2 + \left(\frac{z}{r_{1i}} - \frac{r_{2j}}{r_{1i}} cos() \, sin(\varphi)\right)^2}}. \tag{19}$$

Both coils have one or more turns, and since the equations derived above apply only to the arrangement of simple loops, it is not possible to apply them directly. The calculation is divided into $N_1 \, N_2$ sub-steps, where the mutual inductance of all turn's combinations of the first and second coil is determined. Substituting (17) into (20) we get the total mutual inductance.

$$M = \sum_{i=1}^{N_1} \sum_{j=1}^{N_2} M_{ij} \tag{20}$$

As mentioned above, spiral-shaped coils are rather used for high-frequency applications, while rectangular or square-shaped coils are suitable for applications operating at lower frequencies. On the one hand, there are no electric field gradients, the coupling is rather inductive, and on the other hand we try to make maximum use of the built-up areas to maximize the coupling factor between the coils. As in the previous case, Biot-Savart's law can be applied here as well.

However, since it is not a circular coil, the advantages of the cylindrical coordinate system cannot be used, and the calculation is considerably complicated. To avoid confusing relationships, we will only consider the coaxial arrangement of two coils. These can have different geometries and different numbers of turns.

Figure 10 shows the real and simplified geometry of the coil, on which the derivation of the calculations will be performed. As can be seen in the figure on the left (**Figure 10**), the actual turns have different lengths at the same position, making the whole arrangement asymmetrical. The analytical solution of the field would then be very complicated and quite confusing.

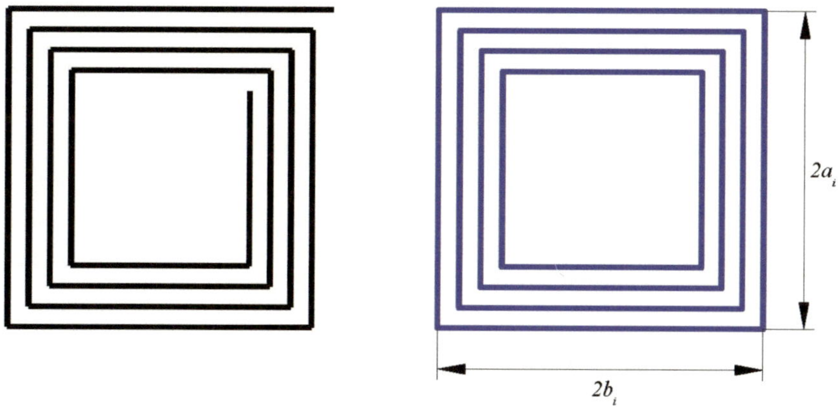

Figure 10.
Real (left) and simplified (right) geometry of the coil with rectangular shape.

Thanks to the equivalent replacement of individual turns with concentric rectangles/squares, we are able to solve the magnetic field around the coil relatively easily and analytically. **Figure 11** indicates the relative position of two coils of different dimensions and number of turns spaced by a length z.

Let us now focus on the i-th turn of the lower coil and the j-th turn of the upper coil. The magnetic field passing through the upper coil (excited by the lower coil) can be calculated from (21), where B_{iz} is the induction of the magnetic field in the z-axis.

$$\phi_{ij} = \int_{S_j} \vec{B}_i d\vec{S}_j = \int_{S_j} B_{iz} dS_j = \int_{S_j} B cos(\theta) dS_j \tag{21}$$

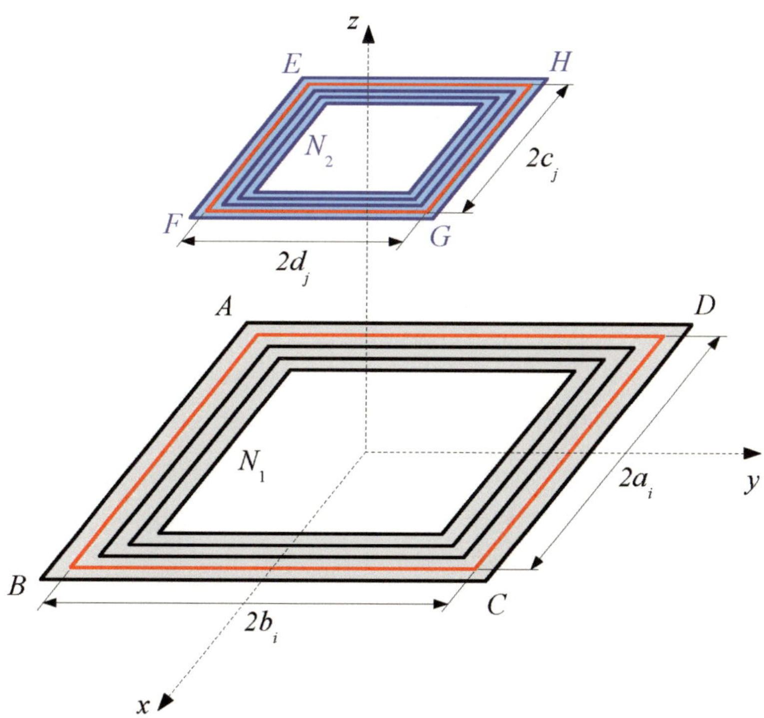

Figure 11.
Simplified situation for determination of mutual inductance between rectangular coils.

Furthermore, we can use Biot-Savart's law to determine the increment of the B_{CD} magnetic field from the current-carrying segment of the i-th turn as:

$$\vec{B}_{CD} = \frac{\mu_0}{4\pi} \frac{1}{\sqrt{(b_i - y)^2 + z^2}} \left[\frac{a_i + x}{\sqrt{(b_i - y)^2 + z^2 + (a_i + x)^2}} + \frac{a_i - x}{\sqrt{(b_i - y)^2 + z^2 + (a_i - x)^2}} \right], \tag{22}$$

An if applicable next equation:

$$cos\,(\theta) = \frac{b_i - y}{\sqrt{(b_i - y)^2 + z^2}}, \tag{23}$$

For z-component of B_{CD} induction we can derive (24)

$$B_{CD-z} = \frac{\mu_0}{4\pi} \frac{b_i - y}{(b_i - y)^2 + z^2} \left[\frac{a_i + x}{\sqrt{(b_i - y)^2 + z^2 + (a_i + x)^2}} + \frac{a_i - x}{\sqrt{(b_i - y)^2 + z^2 + (a_i - x)^2}} \right]. \tag{24}$$

For the total coupled magnetic flux with CD segment, we can write integral in form of (25)

$$\Psi_{CD-z} = \int_{-d_j}^{d_j} dy \int_{-c_j}^{c_j} B_{CD-z} dx. \tag{25}$$

As shown in [24], although the solution of the integral (25) is more complicated, we obtain a purely analytical relation (26).

$$\begin{aligned} \Psi_{CD-z} = \frac{\mu_0}{4\pi} \Bigg[& \Gamma_1 - (a_i + c_j)\, tanh^{-1}\left(\frac{a_i + c_j}{\Gamma_1}\right) - \Gamma_2 \\ & + (a_i - c_j)\, tanh^{-1}\left(\frac{a_i - c_j}{\Gamma_2}\right) - \Gamma_3 \\ & + (a_i + c_j)\, tanh^{-1}\left(\frac{a_i + c_j}{\Gamma_3}\right) - \Gamma_4 \\ & - (a_i - c_j)\, tanh^{-1}\left(\frac{a_i - c_j}{\Gamma_4}\right) - \Gamma_4 \Bigg] \end{aligned} \tag{26}$$

In relation (26) it is still necessary to substitute substitution (27).

$$\begin{aligned} \Gamma_1 &= \sqrt{(b_i + d_j)^2 + z^2 + (a_i + c_j)^2} \\ \Gamma_2 &= \sqrt{(b_i + d_j)^2 + z^2 + (a_i - c_j)^2} \\ \Gamma_3 &= \sqrt{(b_i - d_j)^2 + z^2 + (a_i + c_j)^2} \\ \Gamma_4 &= \sqrt{(b_i - d_j)^2 + z^2 + (a_i - c_j)^2} \end{aligned} \tag{27}$$

The magnetic flux from the other segments (AB, BC and DA) can be easily determined using the same relations. For example, to calculate the segment BC, it is enough to swap a_i with b_i and c_j with d_j in (26). Due to the symmetry, the Eq. (28) will apply.

$$\Psi_{AB-z} = \Psi_{CD-z} a\, \Psi_{DA-z} = \Psi_{BC-z} \tag{28}$$

And because in the case of a unity current considering mutual inductance between the i-th and j-th turns next equation is valid (29)

$$M_{ij}\, \Psi_{ij-z} = \Psi_{AB-z} + \Psi_{BC-z} + \Psi_{CD-z} + \Psi_{DA-z}, \tag{29}$$

the total mutual inductance of both coils is based on (30).

$$M = \sum_{i=1}^{N_1} \sum_{j=1}^{N_2} M_{ij} \tag{30}$$

3.2 Self-inductance of planar coils

To calculate the intrinsic inductance of a planar coil, it is possible to find simple approximation relations, which are suitable for consequent mathematical derivations. However, their big disadvantage is only an approximate calculation with an often-indeterminate error. In addition, the relationships apply only to coils with an equilateral plan

$$L_a = \frac{\mu_0}{2} N^2 D_{AVG} K_1 \left[ln\left(\frac{K_2}{p}\right) + K_3 p + K_4 p^2 \right] \tag{31}$$

Here, for p, the turn's filling factor on the coil surface and D_{AVG} is represented as the mean winding diameter.

$$p = \frac{D_2 - D_1}{D_2 + D_1}, D_{AVG} = \frac{D_2 + D_1}{2} \tag{32}$$

In **Table 3**, the coefficients depending on the approximated coil geometry are calculated (**Figure 12**). The coefficients K_1 - K_4 must always be selected according to the current geometry.

Coil shape	K_1	K_2	K_3	K_4
Circular	1	2.46	0	0.2
Squared	1.27	2.07	0.18	0.13
Hexagonal	1.09	2.23	0	0.17
Octagonal	1.07	2.29	0	0.19

Table 3.
Estimated coefficients for identification of the shape of planar coil.

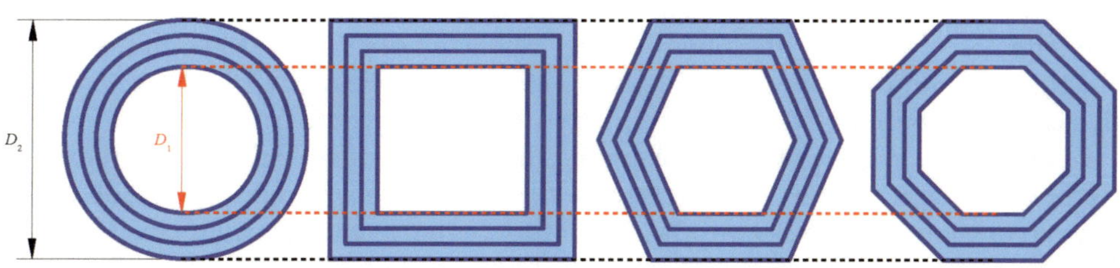

Figure 12.
Allowed degenerations of the coil's geometry for calculation of inductance using Eq. (31).

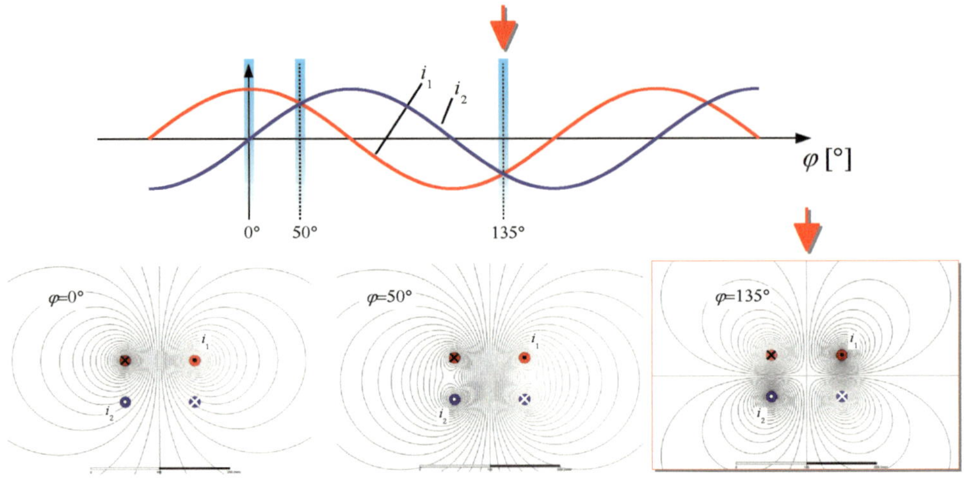

Figure 13.
Time-varying electromagnetic field around the system of coupling coils.

3.3 Coupling coefficient

The magnetic coupling between two coils is formed by a magnetic field, which is generated by a transmitting coil. For many reasons, this array can never be coupled to the receiving coil in its full size, and the larger the array, the better the coupling is achieved. This phenomenon is described by the so-called coupling factor k, which takes values from 0 to 1. The coupling factor depends on the geometry of the coils and their mutual position. Many authors mistakenly qualify the degree of coupling of two coils based on the shape of the electromagnetic field in their surroundings. This approach leads to misinterpretations mainly because the field itself is variable in time and looks different at different times. The situation is clearly shown in **Figure 13**.

For example, in S-S compensation, the currents flowing through the primary and secondary windings are time-shifted by 90 electrical degrees. While at the instant j = 0° and j = 50° the coils appear to be coupled, while at the instant j = 135° they are without mutual coupling according to the shape of the field (**Figure 13**). The only reliable way to determine the coupling factor is to apply relation (33).

$$k = \frac{M}{\sqrt{L_1 L_2}} \tag{33}$$

We will explain some relationships on a simple example, in which we determine the coupling factor of two coaxially placed coils of circular shape with planar design. We will perform the calculation on three similar geometries (**Figure 12** left), where the first pair of coils will have an inner diameter D_1 = 100 mm and an outer diameter D_2 = 200 mm, whereby the other two pairs of coils will have the same dimensions multiplied by two and three. The number of turns is the same N = 5 for all cases. The coupling factor will be plotted for a distance z = 5 ÷ 300 mm.

As shown in **Figure 14** (left), changing the distance of the coils, the coupling factor k decreases rapidly, while the rate of this decrease is highly dependent on the respective geometry of the coils. It is therefore better to choose larger coil dimensions to improve the coupling at higher distances. In **Figure 14** (right) we see the effect of the misalignment of the coils in the x-axis at their constant distance in the z-axis. The coupling factor is somewhat less sensitive to this method of deflection.

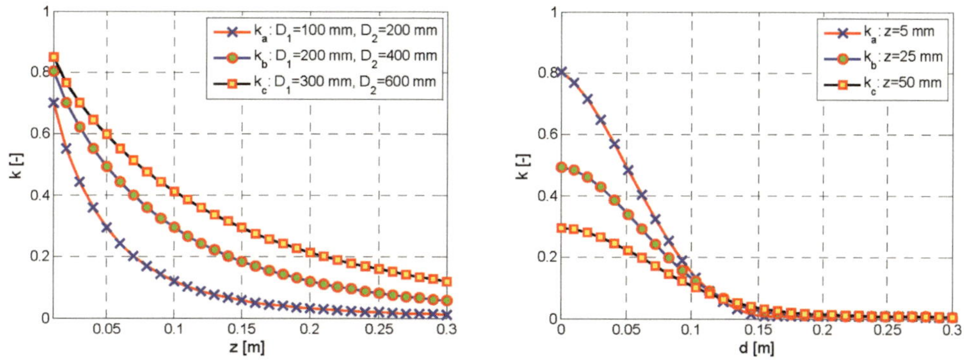

Figure 14.
Dependency of coupling factor on the geometrical properties of the coil and mutual distance (left) and mutual radial displacement (right).

3.4 Parasitic capacitance of the coil

For the capacitance between two turns with mean radius r_i and the distance between individual turns e we can write the relation (34).

$$C_{ip} = \frac{3}{2} \frac{\pi \varepsilon_0 r_i}{ln\left(\frac{e}{r_v}\right)} \tag{34}$$

If we have a coil with N turns, the total parasitic capacitance must be determined as

$$C_{Cp} = \left(\sum_{i=1}^{N-1} \left[\left(\frac{3}{2} \frac{\pi \varepsilon_0 r_i}{ln\left(\frac{e}{r_v}\right)} \right)^{-1} \right] \right)^{-1}. \tag{35}$$

Further to the pattern of **Figure 12** to the left we denote the outer diameter D_2 = 2r2, we get the modification (35) in the form (36).

$$C_{Cp} = \left(\sum_{i=1}^{N-1} \left[\frac{2}{3} \frac{ln\left(\frac{e}{r_v}\right)}{\pi \varepsilon_0 [r_2 - (r_v + e/2)i]} \right] \right)^{-1} \tag{36}$$

The geometric arrangement, according to which (36) can be easily applied, can be seen in **Figure 15**.

Unlike high-frequency systems, at lower operating frequencies, ropes with insulated conductors are used almost exclusively. The reason is the lower influence of parasitic capacitances and especially the better current utilization of the coil.

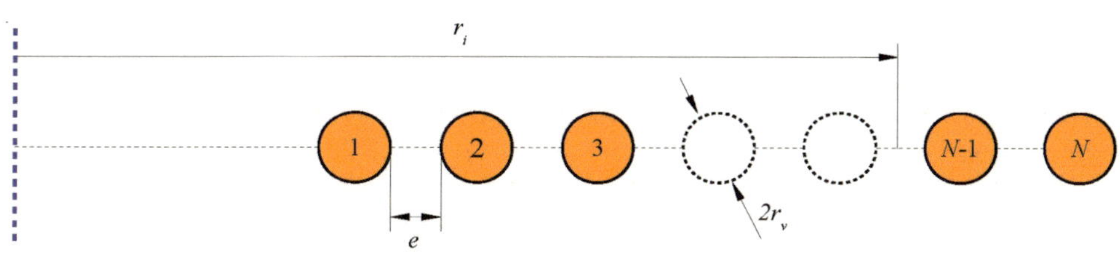

Figure 15.
Situation of the coil's turn placement for the calculation of parasitic capacitance.

3.5 Series parasitic resistance of the coil

The series resistance of the coil is one of the most critical parameters of the system with the greatest influence on its operational efficiency and it is therefore very important to know this value as accurately as possible. We can start with the general relation for resistance according to (37).

$$R = \rho \frac{l}{S} \tag{37}$$

So far, we will not consider temperature or frequency dependences. While the effective area of the conductor S depends only on the current load, the length l already depends on the geometric shape of the coil. As mentioned earlier, spiral planar coils of solid conductor are more suitable for high frequency applications.

3.5.1 Series parasitic resistance of spiral planar coil

In this approximation, we will only talk about coils wound with a copper conductor of circular cross-section. These with their shape most closely resemble parts of the Archimedean spiral, where regarding **Figure 9** on the right we can denote the inner radius as r_{1A} and the outer radius as r_{1B}. The distance between the individual turns e and the number of turns N will be constant. As mentioned earlier, the key length here is played primarily by the length of the conductor of the wound coil. We can write an equation for the Archimedean spiral in polar coordinates

$$r = r_{1A} + \zeta \cdot \varphi, 2\pi\zeta = e. \tag{38}$$

The length of the spiral thus described can be determined by integration (39)

$$l_c = \int_0^{2\pi N} \sqrt{\left(r_{1A} + \frac{e}{2\pi}\varphi\right)^2 + \left(\frac{dr}{d\varphi}\right)^2} d\varphi, \tag{39}$$

However, the disadvantage remains the fact that the integral (39) cannot be solved analytically. It is therefore necessary to integrate numerically for the calculation. It is also possible to use an approximation relation for an approximate calculation

$$l_c = \frac{1}{2}\left[\frac{e}{2\pi}(2\pi N)^2 + 2\pi N\left(\sqrt{r_{1A}^2 + \left(\frac{e}{2\pi}\right)^2} + r_{1A}\right)\right], \tag{40}$$

or simplification by means of an average radius, see (41). The calculation is then very fast and convenient.

$$l_c = 2\pi N \frac{r_{1A} + r_{1B}}{2} \tag{41}$$

In addition, high-frequency applications require winding of a solid conductor to reduce the parasitic capacitance of the coil. Therefore, if we consider the effect of the skin effect, we can adjust (37) to the shape (44) for a conductor with radius r_v. Here for l_c we use one of the equations (39)–(41).

$$R_{c-AC}(\omega) = \begin{cases} \rho_{Cu} \dfrac{l_c}{\pi\delta(2r_v - \delta)}, \wedge \delta \le r_v \\ \rho_{Cu} \dfrac{l_c}{\pi r_v^2}\left(1 + 0.0021\left(\dfrac{r_v}{\delta}\right)^4\right), \wedge \delta > r_v \end{cases}, \delta = \sqrt{\frac{2\rho_{Cu}}{\omega\mu_0}} \tag{42}$$

3.5.2 Series parasitic resistance of rectangular planar coil

The coil has rectangular turns to achieve maximum inductance (**Figure 16**). If we denote the external dimensions of the coil by the letters a and b and consider the spacing between the individual turns *e* constant, we can write Eq. (43) for the resulting resistance.

$$R_{co-DC}(\vartheta) = \frac{\rho(\vartheta)}{S_{Cu}}\left(2\sum_{i=1}^{N}[a+b-[1+(i-1)4]e]\right) \tag{43}$$

Furthermore, if we choose an insulated RF cable with wires whose diameter is much smaller than the penetration depth *d*, we can certainly rule out the effect of the skin effect. We are then talking about *a* conductor with an effective cross section S_{Cu}, whose frequency dependence is caused only by the phenomenon of proximity. With a few modifications, it can be further simplified (44) by removing the summation into the shape

$$R_{co-DC}(\vartheta) = \frac{\rho(\vartheta)}{S_{Cu}}2N[a+b+e(1-2N)] \tag{44}$$

In all the cases described, the turns are evenly distributed in one layer with a constant *e*. The current flowing through the coil thus has the same direction in all turns and generates a magnetic field with lower intensity on the external turns and higher intensity on the internal turns. This magnetic field induces eddy currents into all coil turns and thus increases its overall resistance. This process is commonly referred to as the proximity phenomenon and can be conveniently calculated from the relationship for eddy current losses in individual parallel conductors. For one fiber of diameter d_s of an insulated cable of length *l*, exposed to an external magnetic field *B* of a harmonic waveform of angular frequency *w*, we can write

$$R_{prox} = \frac{\pi l d_s^4}{64\rho}\omega^2 B^2 \tag{45}$$

If we use a cable made of n_f insulated conductors for winding of the coil, it is possible to adjust e_g (44) to the shape (46) by counting (42).

$$R_{cn-AC}(\vartheta,\omega) = \left(\frac{\rho(\vartheta)}{n_s\frac{\pi d_s^2}{4}} + n_s\frac{\pi d_s^4}{64\rho(\vartheta)}\omega^2 B_{st\check{r}}^2\right)nN\left[a + tan\left(\frac{\pi}{n}\right)(1-N)e\right] \tag{46}$$

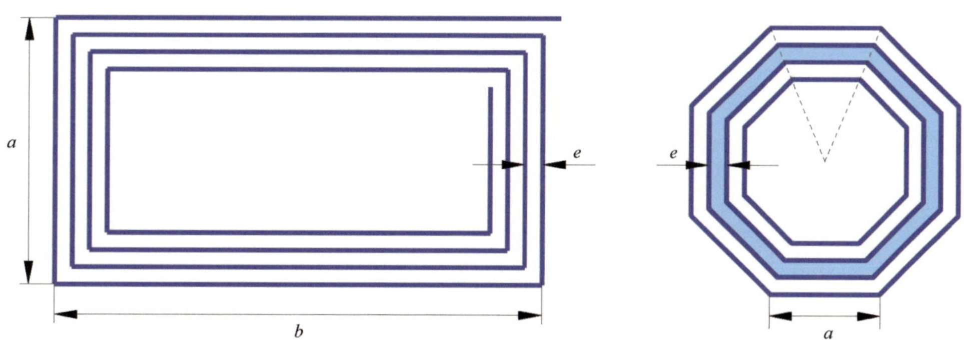

Figure 16.
Rectangular coil identification for the calculation of parasitic resistance.

To illustrate, we will analyze the following geometry. We consider a square coil with an outer edge of length a = 500 mm and N = 20 turns. We choose the spacings between the turn's axes e = 8 mm. The maximum operating frequency is f = 300 kHz, while for a nominal current I = 5 A we choose a current value J = 7 MA/m^2. The calculation is valid for a temperature of 20° C. **Figure 17** shows the dependence of the DC resistance from (44), the AC resistance from (46) and the total resistance. The minimum value of the number of RF wires is determined from (47) and corresponds to the rated current load ($n_{s\text{-}min}$ = 61).

$$n_{s-min} \doteq \frac{4I}{\pi d_s^2 J} \tag{47}$$

By further increase of the number of wires, we therefore only increase the current possibilities of the coil.

As the number of wires increases, the DC resistance R_{DC} decreases sharply, but the AC resistance R_{prox} also increases. The optimum can be found by solving Eq. (48).

$$\frac{d}{dn_s}\left[\left(\frac{\rho(\vartheta)}{n_s \frac{\pi d_s^2}{4}} + n_s \frac{\pi d_s^4}{64\rho(\vartheta)}\omega^2 B_{st\check{r}}^2\right) nN\left[a + tan\left(\frac{\pi}{n}\right)(1-N)e\right]\right] = 0 \tag{48}$$

The condition is met just when it applies

$$n_s \doteq \frac{16\rho(\vartheta)\sqrt{a - e(1+N)\,tan\left(\frac{\pi}{n}\right)}}{\pi\sqrt{\omega^2 B_{st\check{r}}^2 d_s^6\left[a - e(1+N)\,tan\left(\frac{\pi}{n}\right)\right]}}. \tag{49}$$

After substituting, we get the value n_s = 328 wires, which corresponds to the value in **Figure 17**.

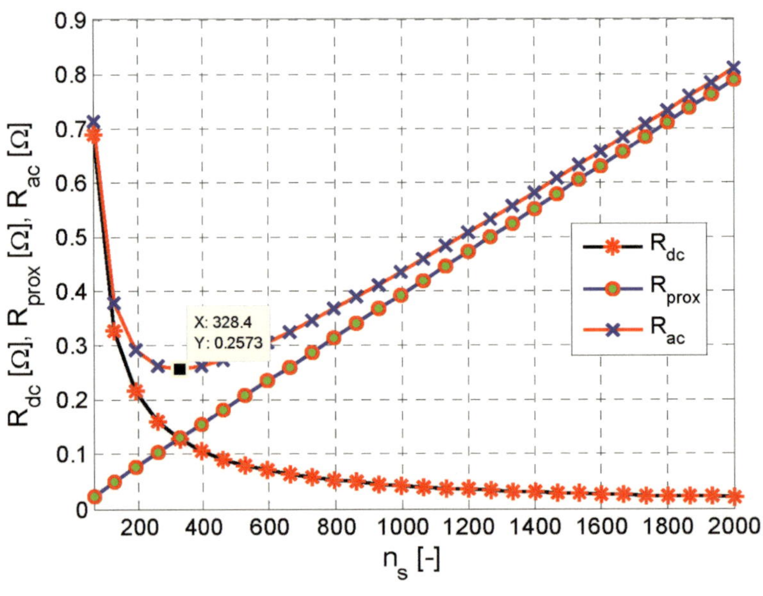

Figure 17.
Nomogram for the calculation of the optimal number of litz wire.

3.5.3 Series parasitic resistance of rectangular planar coil

The most general definition of the quality factor is based on the ratio of accumulated and lost energy in the investigated passive component. For AC supply we can write (50), where the influence of the electric field prevails in the case of a capacitor and the influence of the magnetic field in the case of a coil.

$$Q = \omega \frac{\left|W_{mg} - W_{el}\right|}{P_j} = \frac{1}{R}\sqrt{\frac{L}{C}}. \tag{50}$$

If we consider an ideal coil (R-L circuit) without parasitic capacitance, we get a quality factor such as

$$Q = \omega \frac{\left|W_{mg}\right|}{\mathrm{P}_j} = \omega \frac{\frac{1}{2}LI_m^2}{\frac{1}{2}RI_m^2} = \frac{\omega L}{R}, \tag{51}$$

For more complicated circuits, such as components with parasitic effects, we can also use the relation to calculate the quality factor

$$Q = \left| \frac{I\left\{\frac{\omega L - jR}{\omega C_p (R + j\omega L) - j}\right\}}{R\left\{\frac{\omega L - jR}{\omega C_p (R + j\omega L) - j}\right\}} \right| = \left| \frac{\frac{\omega\left(L - C_p R^2 - \omega^2 C_p L^2\right)}{1 + \omega^2 C_p \left[C_p R^2 + L\left(\omega^2 C_p L - 2\right)\right]}}{\frac{R}{C_p^2 \omega^2 R^2 + \left(\omega^2 C_p L - 1\right)^2}} \right| = \left| \frac{\omega L}{R} - \frac{\omega C_p \left(R^2 + \omega^2 L^2\right)}{R} \right|. \tag{52}$$

The first part of the result of Eq. (52) corresponds to the quality factor of the individual coil, the second part then respects the effect of parasitic capacitance between the turns. A closer look reveals that there is a frequency at which both parts are equal and (52) gives zero result. This frequency is often referred to as the coil's own resonant frequency

$$f_{r-self} = \frac{1}{2\pi\sqrt{LC_p}}. \tag{53}$$

The situation is indicated in **Figure 18** on the right, where the dependency of the quality factor and the character of the resulting reactance on the frequency is plotted for selected values of the parasitic capacity of the inductance and the series

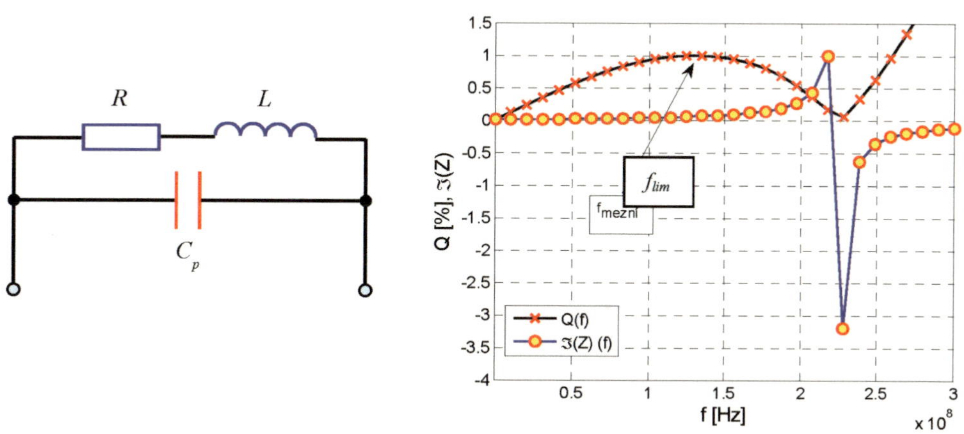

Figure 18.
Parasitic components of the coil and the quality factor characteristic.

resistance of the coil (C_p = 5 pF, L = 0.1 mH, R = 1 Ω). As can be seen from the figure, when reaching the natural resonant frequency of the $f_{r\text{-}self\ circuit}$ »225 MHz, the quality factor is equal to zero and at the same time the inductive character of the reactance changes to capacitive character. For this reason, we always try to operate the coil at a frequency much lower than the self-resonant frequency.

4. Practical design approach for industrial wireless power transfer charging system

4.1 Power electronic system configuration

Electrical engineers responsible for the design of the wireless transfer chargers must consider standard grid network connection during design process. Because many issues are nowadays address on the quality of the supply grid, the main goal during design of any power electronic system is to achieve the best performance related to the power factor parameter at any power consumption of the system. In addition to this fact, it is also required to have fully symmetrical 3-phase current with as low total harmonic distortion as possible [25–28].

$$THDi = \frac{\sqrt{\sum_{\mu=2} i^2_{ac(\mu)}}}{i_{ac(1)}} 100 \tag{54}$$

Regarding above mentioned facts, each power electronic system, which must undergo strict normative given on the qualitative indicators of the grid variables, must be equipped with input active or passive power factor corrector (PFC) and total harmonic distortion correction (THDC). These blocks are consequently followed by diode rectifier, dc/dc converter (step-up or step-down) and the voltage source inverter. Such power electronic system configuration is robust and verified by many similar applications (mostly power supplies and battery chargers). The main negative drawback of such concept lies in higher price and build-in dimensions along with the increase in power rating. This topology should therefore be recommended for low or medium power WPT chargers (**Figure 19** – blue blocks).

Second group of WPT chargers considering the value of power delivery is medium to high power concepts. Here it is recommended to use the configuration composed of input filter (inductive – designed as distribution transformer for

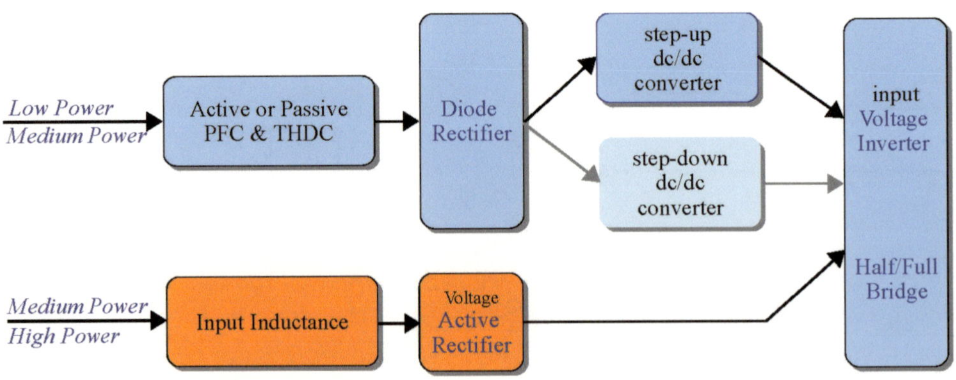

Figure 19.
Power electronics configuration on the primary side of the wireless power charger indicating differences related to the level of the power transfer.

example), followed by the active PFC/THDC rectifier supplying the voltage source inverter (VSI). For both cases (low or high power) the VSI is sourcing primary/transmitting coil with relevant compensation. This configuration of power electronic system (**Figure 19** – orange blocks) is providing low ripple input current with sinusoidal character, low THDi, excellent power factor and controllable output voltage. Therefore, it is not required to implement another dc/dc converter stage within the system [29–31].

The recommended topologies are summarized in **Figure 19** according to system dedicated power level.

The concept of power electronic system for the secondary side also differs based on the type of the load, and level of the power delivery. Basically, it consists of secondary side coil equipped by relevant compensation, passive or active rectifier and dc/dc converter stage providing required functionality of the charger.

Finally, the system connection to the grid considering all the power levels established as WPT categories by SAE TIR J2954 is seen in conceptual layout shown in **Figure 20**, valid especially for central Europe [32–35].

A more detailed example above described solution, which could meet all necessary technical requirements on high power applications and simultaneously having excellent operational properties, is seen in **Figure 21**.

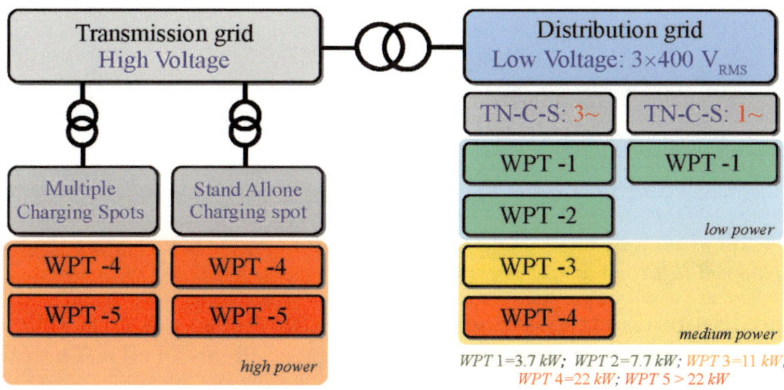

Figure 20.
WPT system categories – Connection to the grid.

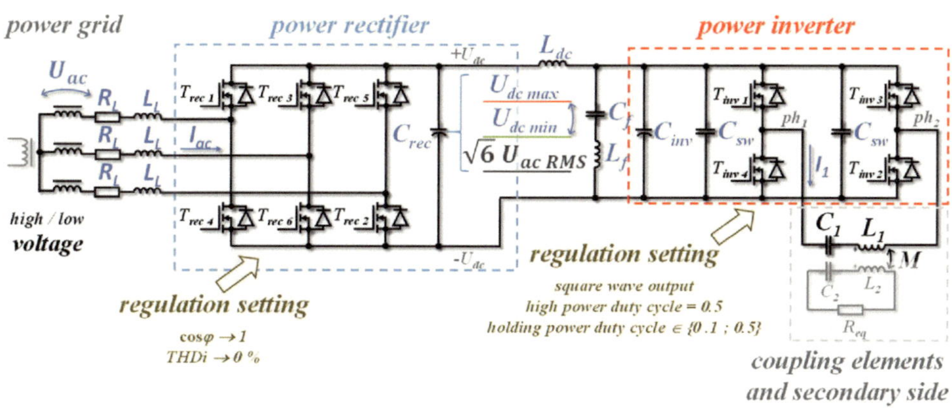

Figure 21.
Recommended system configuration for high power application.

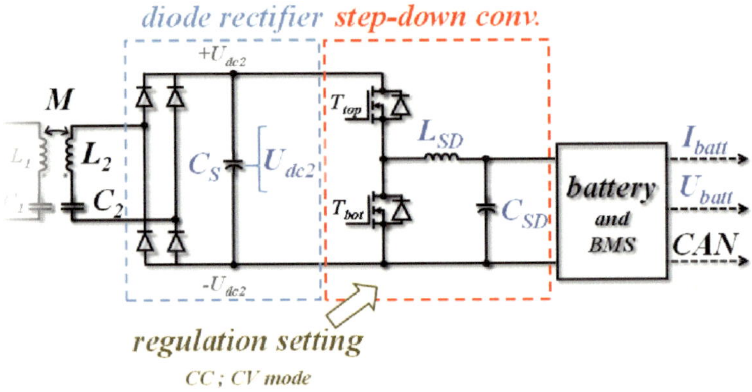

Figure 22.
Recommended system configuration of secondary side for high power application.

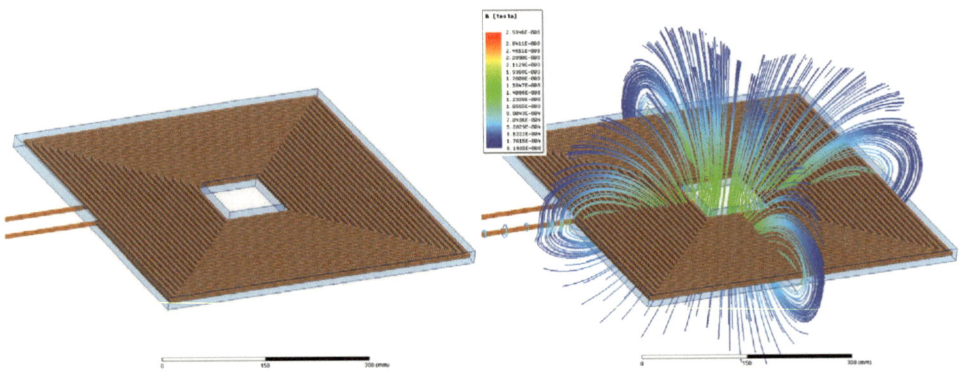

Figure 23.
Proposed coupling coil (left) and its magnetic field (right).

Here the distribution transformer is presented as the grid source, followed by active rectifier, which is responsible for regulation of PF and THDi. Then full bridge inverter is used as VSI and supplies primary side coupling section.

The secondary side of the system shown in **Figure 21** is drawn in more detail in **Figure 22**. The secondary side coupling system is followed by full-bridge diode rectifier with filtering capacitor CS. Then the dc/dc step-down converter (SD) providing required charging algorithm (mostly CC&CV) is supplying the on-board battery pack.

Previously described concepts are representing the mostly used configurations of power electronic systems required for the design of the wireless power chargers suited for industrial and/or automotive applications.

4.2 Coupling elements design

The most important parameter in the design of coupling coils is undoubtedly the product of quality factors *Q* and coupling *k*. Its operating size is strongly dependent on various parameters (e.g. circuit topology, load size, coil distance, etc.) and therefore cannot be optimized directly. One option is to maximize the quality factor. To achieve maximum inductance, we make the coil as planar with square turns (**Figure 23**). In addition, due to the limitation of parasitic capacitance (we now neglect), we keep constant spacings between individual turns of size δ_v = 4 mm. Geometric dimensions allow to wind about 26 turns. In addition, if we know the operating frequency, we can determine the voltage drop and the current through the coil from the required power. It is necessary to design an effective winding cross section for this.

Regarding the available conductor cross-sections, a copper wire (2200 mutually insulated conductors) with a total cross-section of S_v = 19.63 mm2 was selected. The winding produced in this way eliminates the effect of the skin effect and the resulting resistance of the coil is therefore only affected by the phenomenon of proximity.

The coil has 22 turns, the calculation parameters being as follows. The self-inductance has a value L = 147 μH and the active resistance is R = 0.19 Ω.

4.3 Experimental set-up

In this case, we will focus on the WPT 1 category with an output of 3.7 kW. The experimental workplace consists of a programmable power supply, electronic load, precision power analyzer, oscilloscope, input inverter, output rectifier, additional resistors and the compensated LC circuit WPT itself. The measuring workplace is connected according to the functional diagram, see **Figure 24**. The determining factor in the selection of power components was the ability to work with a switching frequency from 200 kHz upwards. For this reason, a solution based entirely on SiC elements was chosen. The inverter is built on 1200 V JFET modules FF45R12J1W1_B11 (Infineon) with a type current of 45 A. Due to the low values of switching times of these modules, which are actually in the order of tens of nano-seconds, it is possible to minimize the effect of inverter dead times. The rectifier is based on a 1200 V diode SiC module APTDC20H1201G (Microsemi) with a type current of 20 A.

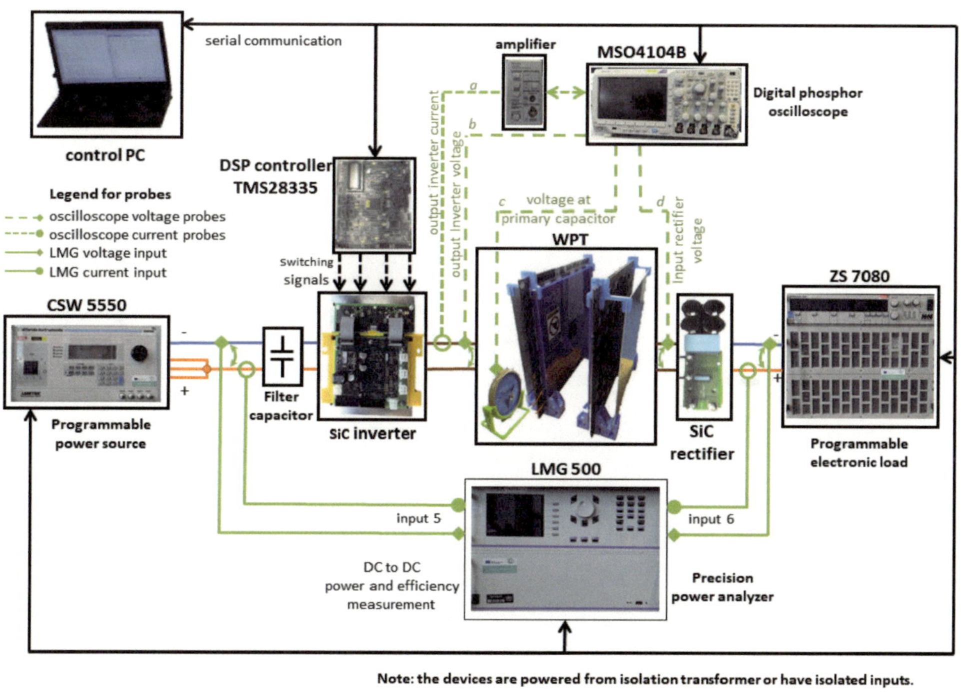

Figure 24.
Block diagram of the laboratory experimental set-up.

4.3.1 Measurement of the operational characteristics of series: series compensated system

On the primary side, a total of three quantities are measured with an oscilloscope. Probe "a" (THDP 0200) measures the output voltage of the inverter, probe "b" (current probe TCP 404 XL and amplifier TCPA 400) measures the primary current and probe "c" (P6015A) senses the voltage on the compensation capacitor. The secondary side is not measured by the oscilloscope at all in this configuration. Also, no resistors are connected here, and the system works directly into the ZS 7080. The applied oscilloscopic measurements on the primary side are rather indicative and do not serve to calculate the efficiency [35–37].

Figure 25 shows an oscillograph at a load power of 2678 W. The purple waveform represents the inverter output voltage, the light blue waveform the primary current waveform, and the blue waveform represents the voltage on the primary compensation capacitor (scale 1: 1000). The real elements (influenced by parasitics) of the WPT system are the main reason why the phase shift of voltage and current is non-zero (according to theoretical assumptions it should be close to zero).

A comparison of power (**Figure 26**) and efficiency (**Figure 27**) shows that the analytical models accurately describe the behavior of the system in a wide range of frequencies and loads.

4.4 Electromagnetic shielding application

Although the current system achieves very high efficiency even over long working distances, it is unsatisfactory due to hygienic limits and standards for EV charging. The main weaknesses are mainly the high switching frequency and the large intensities of the EM field. The magnetic field in the vicinity of both (optimally coupled) coils at a transmitted power of approx. 4000 W is plotted in **Figure 28**. The distribution of the field changes over time, and therefore each time point must be evaluated separately.

The picture shows a large scattering of the field into the surroundings, which must be avoided. Exact induction values at a specific distance from the center of the

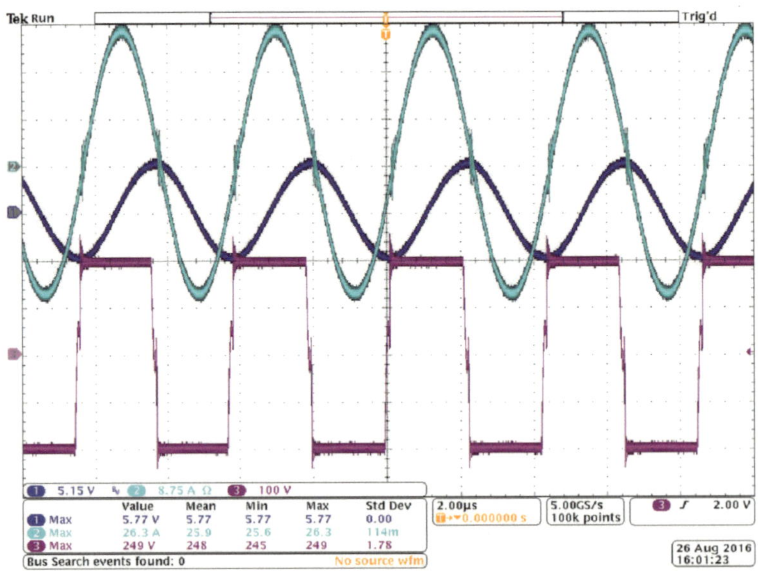

Figure 25.
Time waveforms of the primary side of tested WPT system during full load operation.

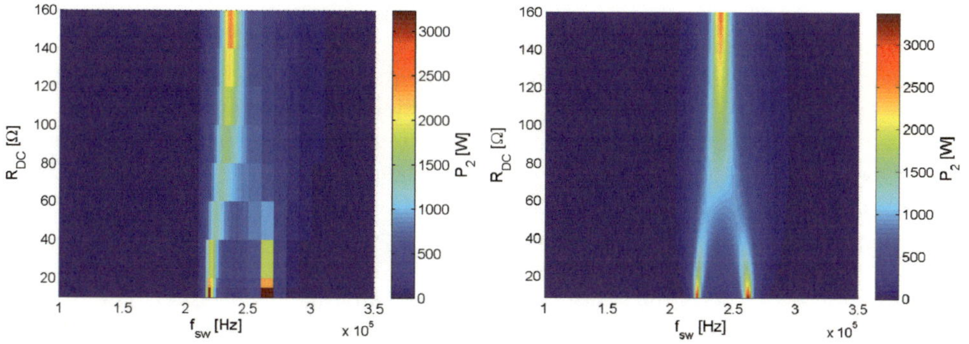

Figure 26.
Output power characteristic in dependency on load and operation frequency for measurement (left) and simulation (right).

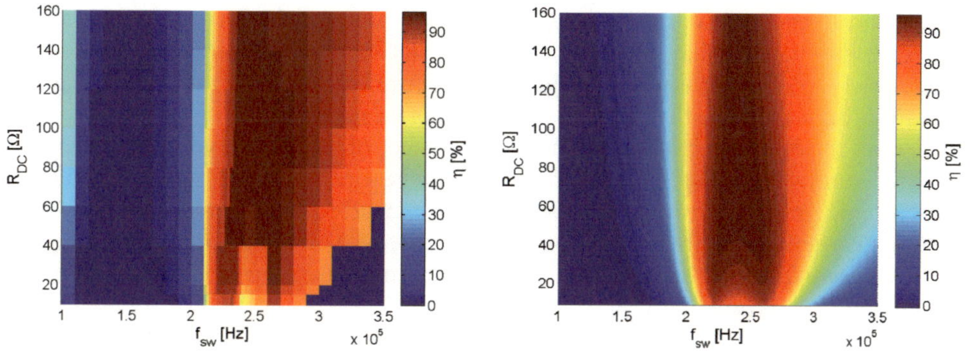

Figure 27.
Efficiency characteristic in dependency on load and operation frequency for measurement (left) and simulation (right).

coils can be obtained by introducing a spherical surface to which the EM field results are mapped. The radius of this area must be defined regarding the dimensions of the vehicle and the location of the coupling coils on its chassis. The key is especially the space in which exposed persons can normally occur. For practical reasons, therefore, it does not make sense to monitor the magnetic induction near both coils. For the sake of clarity, we state here (see **Figure 28**) the magnitude of

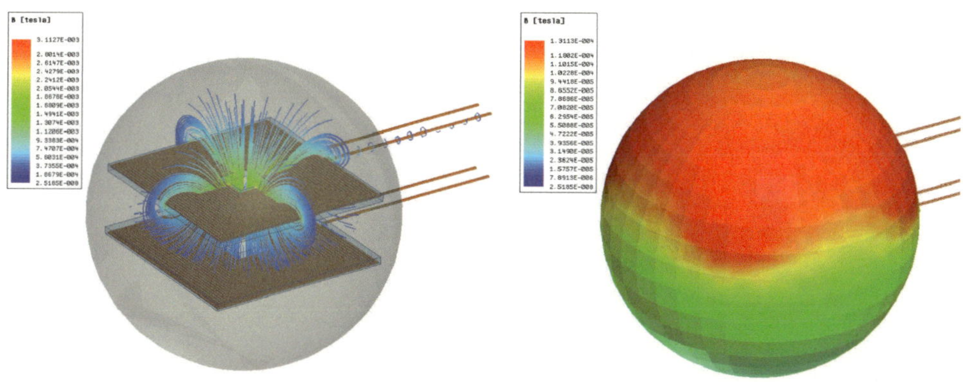

Figure 28.
Magnetic induction around system of unshielded coils.

induction on the sphere surface with a radius of 450 mm at the time (j = 0°), when the current passes through only one coil.

Shielding can be realized by a matrix arrangement of ferrite cores lying on the back sides of both coils. The resulting magnetic field is directed into the main coupling space, while the interior of the vehicle remains protected. The material of the cores must correspond to the operating frequency and especially to the saturation at full load. Material N87 with relative permeability >1450 and operating frequency up to 500 kHz was selected for prototype. The size of the cores is 20x30x3 mm. Due to the high price and weight of the ferrite shield, it is reasonable to lighten its resulting pattern (not to occupy the full area of the coils). The finite element method will be used for this enabling to determine the intrinsic and mutual inductances of coupling coils, ferrite saturation and losses for any arrangement of ferrite cores.

Shielding consists of two functional elements (steps). The first is a ferrite array (plate) that holds the maximum amount of coupled flux and directs it for better bonding to the second coil. The second degree of shielding is an aluminum plate offset over a ferrite field. In the case of supersaturation of the ferrite core, this creates eddy currents that keep the field in the active space of both coils. The situation is indicated in **Figure 29** (left), the ferrite barrier (core) is drawn in gray. The aluminum shield is then shown by a solid plate near the ferrite core.

From **Figure 30** we can see the beneficial effect of shielding even better. Ferrite shielding almost completely shields the field above and below the coils. In this area, the hygienic limits are fully met and without the need for additional shielding.

The magnetic field of the coupling coils (**Figure 30** on the right) is now much better concentrated in the coupling space, which increases the probability of meeting the hygienic limits many times over.

4.4.1 Experimental analysis of the impact of shielding system

In order to verify the theoretical assumptions, an experimental prototype of a previously designed shielding was created. The photograph of the experimental workplace is evident from **Figure 31**.

The aim was to significantly reduce the switching frequency of the supply voltage and to suppress the emission of the EM field to meet the hygienic limits according to "ICNIRP 2010" [38, 39]. The operating parameters of the newly implemented prototype are quantified in **Table 4**. The values are valid for a working distance of 20 cm.

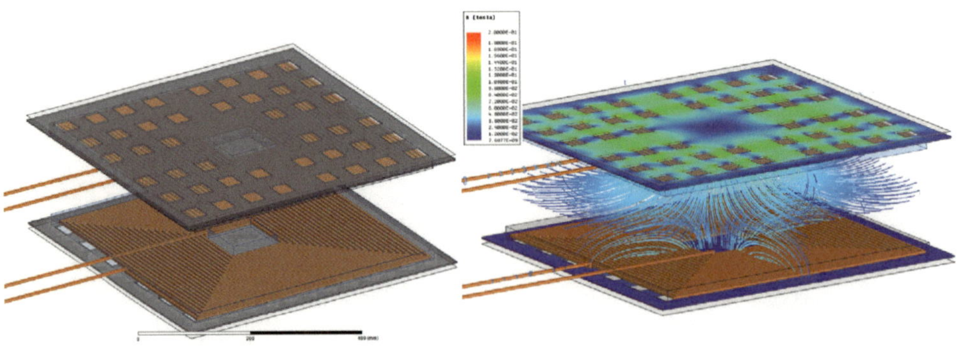

Figure 29.
Proposed electromagnetic shielding (left) and EM field distribution (right).

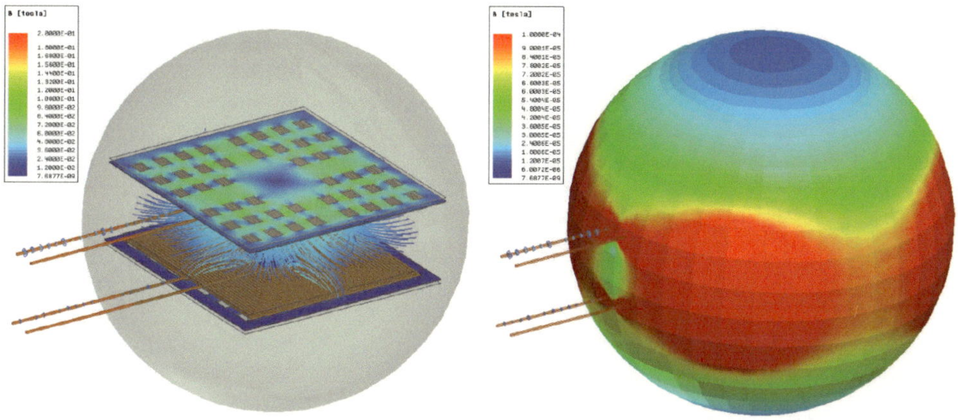

Figure 30.
Magnetic induction around shielded system.

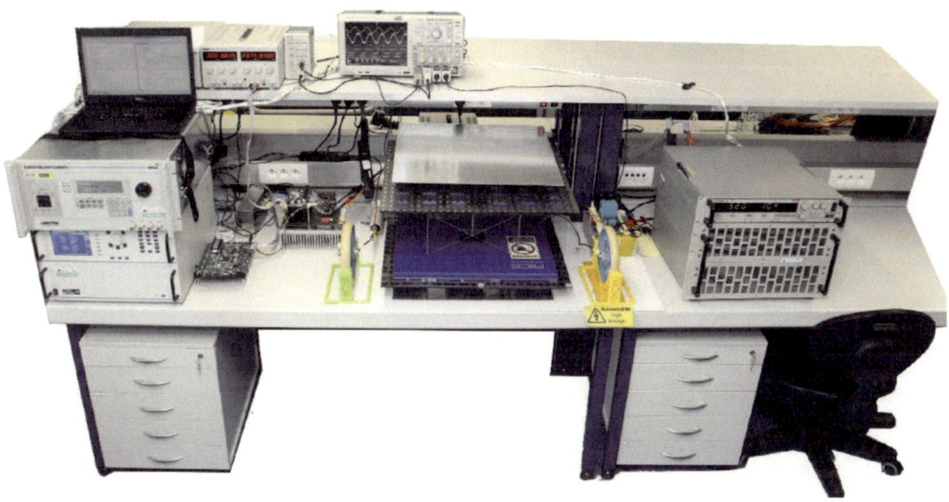

Figure 31.
Laboratory set-up for evaluation of the EM shielding impact.

Optimization to	U_1 [V]	P_1 [W]	P_2 [W]	η [%]	f_r [kHz]	f [kHz]	R_z [Ω]
Power	312	5116	4865	95.1	121.1	121	35
Efficiency	312	4056	3886	95.8	121.1	125	29
		R_1/R_2 [Ω]		L_1/L_2 [μH]		C_1/C_2 [nF]	
		0.21/0.21		172/167		10.444	

Table 4.
Operational parameters of the system after application of the shielding.

Full-scale maps measured at reduced power (maximum efficiency) can be seen in **Figure 32**. The resonance is around 121 kHz, with the high efficiency range more than 10 kHz wide.

The results confirm the ability of the systems to deliver 4 kW to the load at an efficiency of>95%, which, apart from the higher supply frequency, places it in the

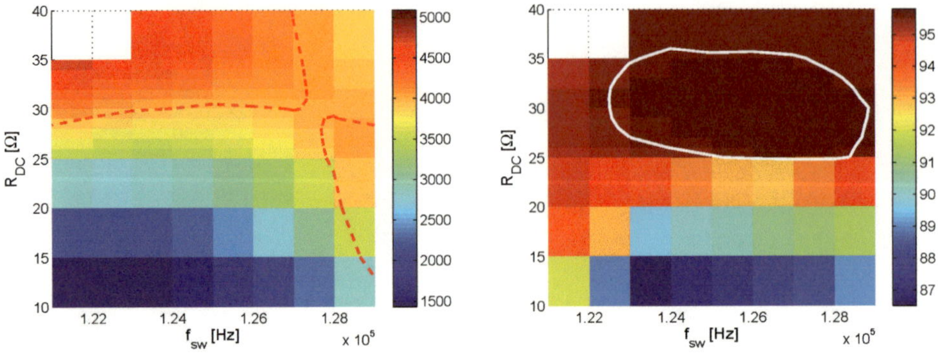

Figure 32.
Output power characteristic (left) and efficiency characteristic (right) for shielded system and 20 cm power transfer distance.

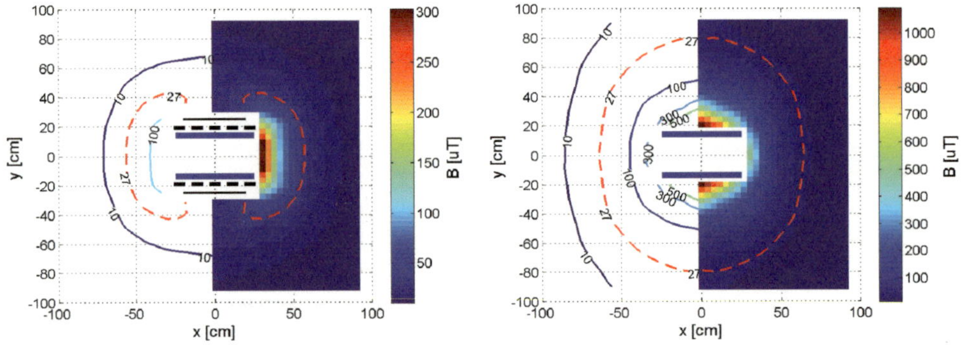

Figure 33.
Evaluation of the values of magnetic field around shielded (system) and non-shielded system (right) during experimental measurement at full power of proposed system.

"WPT 1" category according to the "SAE TIR J2954" wireless charging station standard.

To verify the shielding efficiency, a scattering magnetic field was also measured (measurement uncertainty <2%) around the coupling coils using a calibrated Narda ELT 400 probe. The values were recorded in the cutting plane with a regular step of 10 cm in length (**Figure 33**). The values of the magnetic induction relevant for hygienic limits are boarded by red dashed line (**Figure 33** left). It is seen, that specified limits are achieved approximately 20 cm from the top surface of the coils. Compared to unshielded system (**Figure 33** right), it is reduction of approximately 60 cm considering the spherical distance.

Based on the received and verified results it was achieved, that with the use of presented methodology, it is possible to design wireless charger, whose characteristics will meet standards and normative defined by regulatory companies.

5. Conclusions

The paper has given a brief recapitulation of most important standards and regulations relating to the high-power wireless charging systems. It has proposed the magnetic couplers to be designed exactly according to optimal operation to the specific load.

For medium or high-power wireless chargers, we have recommended to compose the system of input inductance, the active rectifier and the voltage source inverter, which can provide low THDi, excellent power factor and controllable output voltage. Thus, no additional dc/dc converters are needed.

The experimental prototype has proven the validity of presented physical principles and confirmed the proposed conceptual design strategies. It has also shown and discussed the comparison between ac-ac and dc-dc system efficiency relating to losses-to-power transfer ratio.

Additionally, the measurement of leakage magnetic field has shown the real flux density distribution observed around the circular-shaped coupling coils. This could be used for further optimization.

Acknowledgements

This research was supported by project funding APVV – 17 – 0345 - Research of the optimization procedures for improvement of transfer, safety and reliability characteristics of WET systems. This research was also funded by the Ministry of Education, Youth and Sports of the Czech Republic under the project OP VVV Electrical Engineering Technologies with High-Level of Embedded Intelligence CZ.02.1.01/0.0/0.0/18_069/0009855 and by funding program of the University of West Bohemia number SGS-2018-009.

Conflict of interest

The authors declare no conflict of interest.

Author details

Vladimir Kindl[1], Michal Frivaldsky[2]*, Jakub Skorvaga[2] and Martin Zavrel[1]

1 Faculty of Electrical Engineering, University of West Bohemia, Pilsen, Czech Republic

2 Faculty of Electrical Engineering and Information Technologies, University of Zilina, Zilina, Slovakia

*Address all correspondence to: michal.frivaldsky@feit.uniza.sk

References

[1] Zarikas, Vasilios, and Nick Papanikolaou. Advanced and Intelligent Control in Power Electronics and Drives. Auris Reference, 2018.

[2] Li, Y.; Mai, R.; Lu, L.; Lin, T.; Liu, Y.; He, Z. Analysis and Transmitter Currents Decomposition Based Control for Multiple Overlapped Transmitters Based WPT Systems Considering Cross Couplings. IEEE Trans. Power Electron. 2018, 33, 1829–1842.

[3] F. Lu, H. Zhang, H. Hofmann and C. C. Mi, "An Inductive and Capacitive Combined Wireless Power Transfer System WithLC-Compensated Topology," in IEEE Transactions on Power Electronics, vol. 31, no. 12, pp. 8471–8482, Dec. 2016. doi: 10.1109/TPEL.2016.2519903

[4] M. Heidarian, S. J. Burgess, R. Prabhu and N. Fough, "Maximising Inductive Power Transmission using a Novel Analytical Coil Design Approach," *2019 IEEE Wireless Power Transfer Conference (WPTC)*, London, United Kingdom, 2019, pp. 158–161, doi: 10.1109/WPTC45513.2019.9055553.

[5] W. Zhang and C. C. Mi, "Compensation Topologies of High-Power Wireless Power Transfer Systems," in IEEE Transactions on Vehicular Technology, vol. 65, no. 6, pp. 4768–4778, June 2016. doi: 10.1109/TVT.2015.2454292

[6] I. Sîrbu and L. Mandache, "Comparative analysis of different topologies for wireless power transfer systems," 2017 IEEE International Conference on Environment and Electrical Engineering and 2017 IEEE Industrial and Commercial Power Systems Europe (EEEIC/I&CPS Europe), Milan, 2017, pp. 1–6. doi: 10.1109/EEEIC.2017.7977863

[7] Babaki, Amir et al. "Performance Optimization of Dynamic Wireless EV Charger Under Varying Driving Conditions Without Resonant Information." IEEE Transactions on Vehicular Technology 68 (2019): 10429–10438, DOI: 10.1109/TVT.2019.2944153

[8] Y. Jiang, L. Wang, Y. Wang, J. Liu, X. Li and G. Ning, "Analysis, Design, and Implementation of Accurate ZVS Angle Control for EV Battery Charging in Wireless High-Power Transfer," in IEEE Transactions on Industrial Electronics, vol. 66, no. 5, pp. 4075–4085, May 2019. doi: 10.1109/TIE.2018.2795523

[9] Y. Bezawada, R. Fu and Y. Zhang, "Impacts of Coupling Plates on Single-Switch Capacitive-Coupled WPT Systems," *2019 IEEE PELS Workshop on Emerging Technologies: Wireless Power Transfer (WoW)*, London, United Kingdom, 2019, pp. 330–334, doi: 10.1109/WoW45936.2019.9030651.

[10] Kang Hyun Yi, "High frequency capacitive coupling wireless power transfer using glass dielectric layers," *2016 IEEE Wireless Power Transfer Conference (WPTC)*, Aveiro, 2016, pp. 1–3, doi: 10.1109/WPT.2016.7498857.

[11] Chieh-Kai Chang, G. G. Da Silva, A. Kumar, S. Pervaiz and K. K. Afridi, "30 W capacitive wireless power transfer system with 5.8 pF coupling capacitance," *2015 IEEE Wireless Power Transfer Conference (WPTC)*, Boulder, CO, 2015, pp. 1–4, doi: 10.1109/WPT.2015.7140184.

[12] IEEE Standard for Safety Levels with Respect to Human Exposure to Radio Frequency Electromagnetic Fields, 3 kHz to 300 GHz Amendment 1: Specifies Ceiling Limits for Induced and Contact Current, Clarifies Distinctions between Localized Exposure and Spatial Peak Power Density," in IEEE Std C95.1a-2010 (Amendment to IEEE Std C95.1–2005), vol., no., pp.1–9, 16

March 2010 doi: 10.1109/IEEESTD.2010.5433227

[13] J. Kim et al., "Coil Design and Shielding Methods for a Magnetic Resonant Wireless Power Transfer System," in Proceedings of the IEEE, vol. 101, no. 6, pp. 1332–1342, June 2013. doi: 10.1109/JPROC.2013.2247551

[14] Q. Zhu, Y. Zhang, Y. Guo, C. Liao, L. Wang and L. Wang, "Null-Coupled Electromagnetic Field Canceling Coil for Wireless Power Transfer System," in IEEE Transactions on Transportation Electrification, vol. 3, no. 2, pp. 464–473, June 2017. doi: 10.1109/TTE.2016.2633798

[15] J. Kim, H. Kim, C. Song, I. Kim, Y. Kim and J. Kim, "Electromagnetic interference and radiation from wireless power transfer systems," 2014 IEEE International Symposium on Electromagnetic Compatibility (EMC), Raleigh, NC, 2014, pp. 171–176. doi: 10.1109/ISEMC.2014.6898964

[16] Kindl, V., Frivaldsky, M., Spanik, P., Piri, M. and Jaros, V. (2017), "Transfer properties of various compensation techniques for wireless power transfer system including parasitic effects", COMPEL - The international journal for computation and mathematics in electrical and electronic engineering, Vol. 36 No. 4, pp. 1198–1219. https://doi.org/10.1108/COMPEL-04-2016-0143

[17] Y. Wang, Y. Yao, X. Liu, D. Xu and L. Cai, "An LC/S Compensation Topology and Coil Design Technique for Wireless Power Transfer," in *IEEE Transactions on Power Electronics*, vol. 33, no. 3, pp. 2007–2025, March 2018, doi: 10.1109/TPEL.2017.2698002.

[18] H. Zeng, S. Yang and F. Z. Peng, "Design Consideration and Comparison of Wireless Power Transfer via Harmonic Current for PHEV and EV Wireless Charging," in IEEE Transactions on Power Electronics, vol. 32, no. 8, pp. 5943–5952, Aug. 2017. doi: 10.1109/TPEL.2016.2616111

[19] P. Spanik, M. Frivaldsky, M. Piri and V. Kindl, "Wireless power transfer system with reduced voltage stress on compensation capacitors," IECON 2016 - 42nd Annual Conference of the IEEE Industrial Electronics Society, Florence, 2016, pp. 1190–1195. doi: 10.1109/IECON.2016.7793103

[20] L. Ning, Y. Xiao and Z. Ning, "Design of transcutaneous coupling wireless charger," 2011 6th International Conference on Computer Science & Education (ICCSE), Singapore, 2011, pp. 41–46, DOI: 10.1109/ICCSE.2011.6028581

[21] Z. Zhu, C. Zhang and D. Li, "A Novel Parameter Design and Optimization Method for Wireless Power Transfer System," *2018 Asian Conference on Energy, Power and Transportation Electrification (ACEPT)*, Singapore, 2018, pp. 1–5, doi: 10.1109/ACEPT.2018.8610852.

[22] Babic, S. I., Sirois, F., Akyel, C.; "Validity Check of Mutual Inductance Formulas for Circular Filaments With Lateral and Angular Misalignments," Progress In Electromagnetics Research M, Vol. 8, 15–26, 2009

[23] Babic, S. I., Akyel, C.; "Calculating Mutual Inductance Between Circular Coils With Inclined Axes in Air," IEEE TRANSACTIONS ON MAGNETICS, VOL. 44, NO. 7, JULY 2008

[24] S. Kim, H. Park, J. Kim, J. Kim and S. Ahn, "Design and Analysis of a Resonant Reactive Shield for a Wireless Power Electric Vehicle," in IEEE Transactions on Microwave Theory and Techniques, vol. 62, no. 4, pp. 1057–1066, April 2014. doi: 10.1109/TMTT.2014.2305404

[25] T. Shijo et al., "85 kHz band 44 kW wireless power transfer system for rapid

contactless charging of electric bus," 2016 International Symposium on Antennas and Propagation (ISAP), Okinawa, 2016, pp. 38–39.

[26] Bausiere, Robert, et al. Power Electronic Converters DC-DC Conversion. Springer Berlin, 2013.

[27] Y. Yang, W. Zhong, S. Kiratipongvoot, S. Tan and S. Y. R. Hui, "Dynamic Improvement of Series–Series Compensated Wireless Power Transfer Systems Using Discrete Sliding Mode Control," in *IEEE Transactions on Power Electronics*, vol. 33, no. 7, pp. 6351–6360, July 2018, doi: 10.1109/ TPEL.2017.2747139.

[28] F. Genco, M. Longo, P. Livreri and A. Trivino, "Wireless Power Transfer System Stability Analysis for E-bikes Application," *2019 AEIT International Conference of Electrical and Electronic Technologies for Automotive (AEIT AUTOMOTIVE)*, Torino, Italy, 2019, pp. 1–5, doi: 10.23919/ EETA.2019.8804499.

[29] S. Jayalath and A. Khan, "Analysis of the Relationship between the Parameters of IPT Transformer and Power Electronic System," *2018 IEEE Wireless Power Transfer Conference (WPTC)*, Montreal, QC, Canada, 2018, pp. 1–4, doi: 10.1109/ WPT.2018.8639141.

[30] W. Peng and Z. Chen, "Enhanced Planar Wireless Power Transfer Systems with Ferrite Material," *2018 IEEE Wireless Power Transfer Conference (WPTC)*, Montreal, QC, Canada, 2018, pp. 1–4, doi: 10.1109/ WPT.2018.8639455.

[31] J. Pries, V. P. N. Galigekere, O. C. Onar and G. Su, "A 50-kW Three-Phase Wireless Power Transfer System Using Bipolar Windings and Series Resonant Networks for Rotating Magnetic Fields," in IEEE Transactions on Power Electronics, vol. 35, no. 5, pp. 4500–4517, May 2020, doi: 10.1109/ TPEL.2019.2942065.

[32] V. Kindl, M. Frivaldsky, M. Zavrel, and M. Pavelek, "Generalized Design Approach on Industrial Wireless Chargers", Energies, vol. 13, no. 11, 2697, 2020.

[33] Kalialakis, C., & Georgiadis, A. (2014). The regulatory framework for wireless power transfer systems. Wireless Power Transfer, 1(2), 108–118. doi:10.1017/wpt.2014.13

[34] B. M. Mosammam, N. Rasekh, M. Mirsalim and A. Khorsandi, "Electromagnetic Analysis for DD Pad Magnetic structure of a Wireless Power Transfer (WPT) for Electrical Vehicles," *2018 Smart Grid Conference (SGC)*, Sanandaj, Iran, 2018, pp. 1–6, doi: 10.1109/SGC.2018.8777750.

[35] J. Chakarothai, K. Wake, T. Arima, S. Watanabe and T. Uno, "Exposure Evaluation of an Actual Wireless Power Transfer System for an Electric Vehicle With Near-Field Measurement," in IEEE Transactions on Microwave Theory and Techniques, vol. 66, no. 3, pp. 1543–1552, March 2018. doi: 10.1109/ TMTT.2017.2748949

[36] ICNIRP2010. International Commission on Non-Ionizing Radiation Protection. 2010. Available online: h ttp://www.icnirp.org/cms/upload/pub lications/ICNIRPLFgdl.pdf (accessed on 7 August 2018).

[37] H. Kim et al., "Coil Design and Measurements of Automotive Magnetic Resonant Wireless Charging System for High-Efficiency and Low Magnetic Field Leakage," in IEEE Transactions on Microwave Theory and Techniques, vol. 64, no. 2, pp. 383–400, Feb. 2016. doi: 10.1109/TMTT.2015.2513394

[38] SAE TIR J2954. SAE International. 2016. Available online: http://standards.

sae.org/wip/j2954/ (accessed on 19 January 2017).

[39] T. Iwamoto et al., "Measurement of electromagnetic field in the vicinity of wireless power transfer system for evaluation of human-body exposure," 2014 International Symposium on Electromagnetic Compatibility, Tokyo, Tokyo, 2014, pp. 529–532.

Section 3

Wireless Power Transfer in Communication Systems

Chapter 5

Convergence of Wireless and Optical Network in Future Communication Network

Rajarshi Mahapatra

Abstract

The requirement of data increases many-fold in recent years to support the newest technologies in B5G and 6G. Wireless is the last mile solution as access with an optical network as the backbone in future communication systems. Over the years in every new generation, the distance between the base station and the user is decreasing and the optical node is coming closer to the user. There are several technologies like AR/VR, AI, holographic communication, holographic telepresence, etc. are the main candidates in B5G and 6G, which are required high-speed connection with low latency. To support these services, it is almost mandatory that transmit data across the network should be smooth and seamless to provide successful communication. Providing a successful and appropriate wireless link among the users simultaneously to achieve the requirements is becoming more complex, hence challenging. The optical backbone of all wireless access networks requires supporting these user's requirements, needs to evolve continuously with wireless network evolution. This chapter will study the evolution of both networks to understand their cooperation, alignment, and support.

Keywords: Wireless network, 5G, Heterogeneous network, Massive MIMO, 6G, Cognitive radio, Optical network, Cognitive optical network, Ethernet PON

1. Introduction

Over the year, the number of connected devices (wirelessly and wired) is ever-increasing, will reach 13 billion by 2023 [1]. To support these always-connected devices, the demand for high speed, high reliability, low-latency, low-cost, dense connectivity, different types of mobility needs, and heterogeneous connectivity is escalating, which forced the telecommunications industry to enter into a new era of the future communication network (FCN) [2]. Furthermore, to unleash the full potential of Industry 4.0, guaranteed real-time communication between humans, robots, factory logistics, and products is a fundamental requirement [3]. The FCN incorporates 5G and beyond 5G network, whose main objectives will be application/service-oriented, which are on-demand and highly heterogeneous in nature [4]. To support ever-increasing devices for application-specific on-demand services, there is a strong requirement to view, design, and optimize the network from an end-to-end perspective.

To supports ever-increasing demand on requirements for different types of usage, applications, services, several technologies have been developed over the year. **Table 1** describes the important milestones in both wireline and wireless communication. The development of wireline communication first started in copper and later shifted to the optical domain. In the present day, optical fiber is used in the backhaul network and copper wire is used normally in the access network. In the case of wireless communication, communication first started in the sub-GHz range and slowly it moves towards high-frequency ranges. In the latest, wireless communication is moving towards the 60–100 GHz range (mmWave communication) [4, 5].

To facilitate 5G capabilities (latency less than 1 ms, more than 5 Gbps data rate for high mobile user, other quality of (QoS) and quality of experience (QoE), enhanced spectral, energy and network efficiency, smart security, etc.) FCNs need to enhance existing services. To fulfill 5G and beyond 5G stringent service requirements, it is essential to have an understanding of all available resources across networks (wireless and optical), across radio-access technologies (RAT) (various frequency domain), across services (different class of services and traffic type), across emerging and disruptive technologies (internet-of-things (IoT), artificial intelligence (AI), augmented reality/virtual reality (AR/VR)), and across cloud

Wireline Communication		**Wireless Communication**	
Year	**Milestones**	**Year**	**Milestones**
1876	A. G. Bell transmits the first sentence	1894	Transmission through radio demonstrated by J. C. Bose
1877	First long-distance telephone line	1986	Marconi demonstrates wireless telegraphy
1927	The first transatlantic phone call, from the US to the UK	1901	Send the signal wirelessly across the Atlantic
1948	Shannon published Shannon's formula	1914	first voice communication was established over a radio
1956	Kapany invented the glass-coated glass rod, named Fiber	1946	The first public mobile telephone was introduced by AT&T
1958	LASER invented by Schawlow and Townes	1973	Motorola makes a mobile call from a handheld mobile phone
1960	Kao demonstrate communication through fiber	1992	GSM starts its operation
1970	Corning Glass produced a practical fiber	1997	IEEE releases WiFi standard
1973	TCP/IP protocol proposed by Kahn and Cerf	2003	Birth of WWWW
1977	the first live telephone traffic through fiber optics	2009	Birth of the Internet of Things
1992	Birth of WWW	2010	First 4G handset introduced
1997	Fiber optic link around the globe	2012	5G focus group created
2005	YouTube.com launches	2016	Coined Industry 4.0
2006	Cloud computing started	2016	Google unveils Google Assistant
2014	Demonstration of software-defined networking	2019	6G Communication coined

Table 1.
Important milestones in wireless and wireline communication.

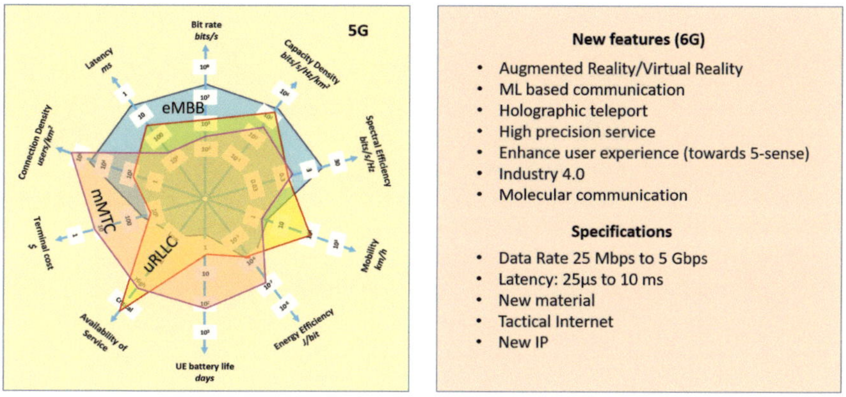

(a) (b)

Figure 1.
Characteristics of (a) 5G and (b) 6G communication.

domains, and finally different backhaul network technologies. The 5G applications categories into three main domains: ultra-reliable low latency communication (uRLLC), massive machine type communication (mMTC), and enhanced mobile broadband (eMBB) [6, 7]. Moving beyond 5G, 6G communication includes few disrupting technologies, such as machine learning (ML) based communication, augmented reality/virtual reality (AR/VR), holographic communication, high precision service, enhance user experience (towards 5-sense), Industry 4.0, molecular communication and more. Their specifications are futuristic, which include 5 Gbps in data rate, 25 μs in latency, new material for 5-sense experience, etc. **Figure 1** provides an overview of the required specifications for three different areas in 5G communication and also in 6G communication [4].

2. Technology evolution over the years

In today's telecommunication world, user access the services through different transmission media (copper, wireless, and fiber), however, backbone are predominantly optical. Most of the time, the access network is wireless, as the number of devices increases over the years due invent of IoT). In this work, users use the wireless networks for access purposes with the backbone network as optical. **Figure 2** gives an idea of how the evolution of optical networks makes an impact on the wireless network. As the requirements of high data rate and low latency are increasing, the availability of optical networks (fronthaul) is coming closer to the home and access distance through wireless is decreasing. In the following, the development of technologies will be discussed in both domains.

2.1 Development of wireless network

Over the years, wireless communication evolved generation-wise, started from 1G analog to 5G digital and moving towards 6G communication. The focus of 5G and 6G technologies is to connect people, society seamlessly along with applications, services, data, and geographical area in a smart networked environment. The present wireless network is heterogeneous in terms of infrastructure (Macrocell to femtocell), spectrum usage (licensed and unlicensed, sub-GHz to THz), coverage (multi-tier), antenna (single to the massive number of antennas), cooperation (user to eNB), and power usage (mW to 100 W).

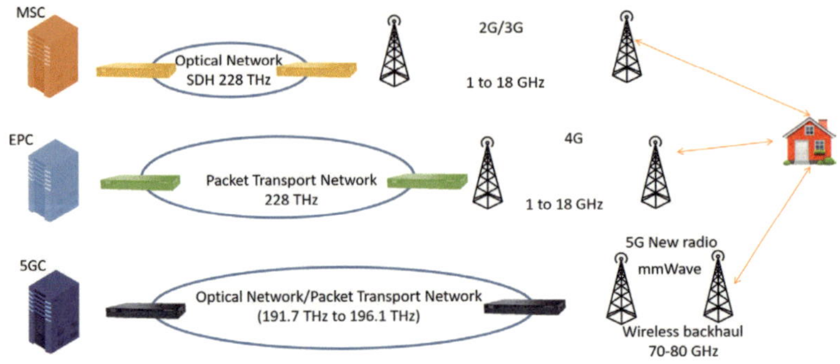

Figure 2.
Impact of the evolution of technologies.

The technologies developed for supporting these heterogeneous characteristics are co-existing together. These technologies are used to serve their purpose and produce interference on other services while in use, due to this their performance is somehow limited. To enhance performance by increasing awareness and cooperation, 5G technologies proposed several new solutions. These include technologies (as shown in **Figure 3**) like massive multiple-input multiple-output (MIMO) for higher data rate and better coverage, coordinated multipoint transmission (CoMP) for a lower outage, distributed antenna system (DAS) for better connectivity, software-defined radio (SDR) for reconfigurability, cognitive radio (CR) for better spectrum utilization, cloud computing for better usage, software-defined network (SDN) for an optimized network, and mmWave communication for high bandwidth. These technologies differ in channel characteristics, usage specification, operational requirement, application supports, etc. The 5G communication stipulates to support a data rate of more than 5 Gbps, less than 1 ms latency for high mobility users [5]. Several important developments in 5G wireless networks are.

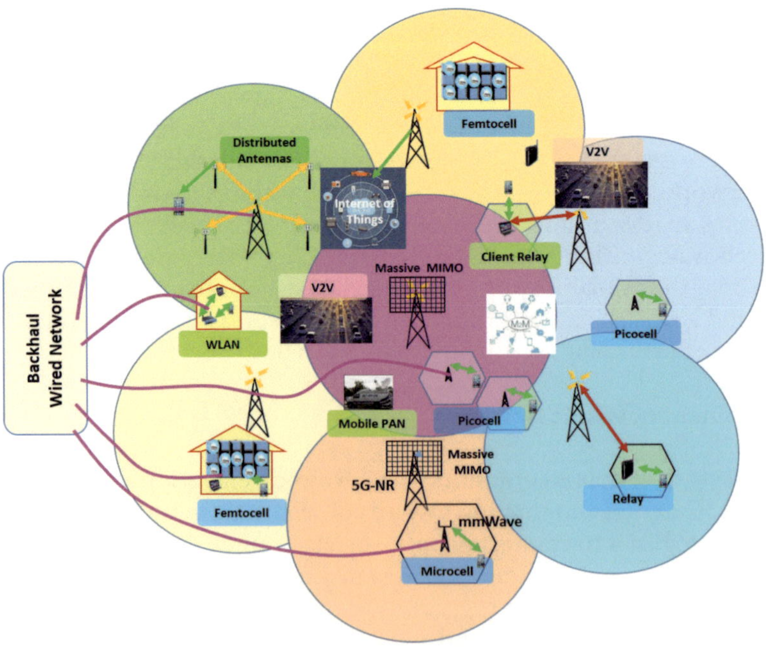

Figure 3.
A typical scenario of a heterogeneous wireless network.

5G New Radio: Even the existence of various radio technologies, 5G communications proposed a completely new radio interface, named 5G new radio (5G NR). The 5G NR interface is a flexible air interface that supports the mainly three ITU defined categories: uRLLC, mMTC, and eMBB. It can also support various other 5G applications such as automotive and health care. 3GPP defined two frequency ranges: FR1 (below 6 GHz) and FR2 (above 24 GHz) [7, 8].

Massive MIMO: The 5G communication uses massive MIMO as a promising multi-user MIMO technology, where the number of antennas (more than 100) at eNB is much more compared to traditional systems. This massive number of antennas allows substantial gains in system capacity and energy efficiency of both users and the system. Due to the increasing number of antennas, which guides to a more spatial resolution; subsequently, several users can use the same time-frequency resource. This can eventually lead to large capacity gains [9].

Non Orthogonal Multiple Access (NOMA): Over the year, orthogonal frequency division multiplexing (OFDM) is the most preferred transmission technique. However, a non-orthogonal scheme (NOMA) has been proposed for efficient 5G communication. In NOMA, each user are distinguished by their power levels while operating in the same band and at the same time. It works with successive interference cancelation (SIC) at the receiver uses and with the help of superposition coding at the transmitter, all users can utilize the same and entire spectrum band. The transmitter site superimposed all the individual signals into a single waveform, while the receiver finds the desired signal with the help of SIC decodes mechanism [10, 11].

mmWave Communication: Availability of large bandwidth in the millimeter range, 5G & beyond wireless systems proposed to use mmWave communications. The mmWave cites a very short wavelength of the radio frequency spectrum between 24GHz and 100GHz. Due to the much shorter wavelength at millimeter band, it allows the deployment of massive antennas at the transceiver. Thus, the large propagating attenuation due to high frequency will be compensated by using a large antenna array, which provides high gains and finally, provides faster data speeds. In dense deployments scenario, it is also suitable for efficient and flexible wireless backhauling, in addition to supporting ultra-high-speed radio access [12].

Internet of Things (IoT): In the present day, IoT is used almost every possible scenario and application. IoT interconnects different types of devices for various applications and enables machine-to-machine (M2M) communication. BY doing so, it enables data communication between heterogeneous devices automatically without human monitoring and control intervention. Several wireless technologies along with few open standards (Vodafon's Cellular IoT and the NB-IoT by 3GPP) have been used for the deployment of IoT. The 5G will be able to provide a connection to a massive IoT network, where billions of smart devices can be connected to the Internet. Since the 5G networks provide flexible and faster networks, IoT can be easily integrated with the wireless software define networking (WSDN) paradigm [13].

Coordinated Multipoint (CoMP): 5G communication supports small cell and the availability of numerous devices in the environment, makes the network very dense. In the dense environment, intercell interference will be more severe for edge users, which is one of the main reasons for the repeated outage. CoMP transmission technique exploits this interference scenario to enhance the users' performances. CoMP mechanism utilizes the resources more effectively and efficiently by dynamic coordination or transmission and reception with multiple eNBs, which eventually improves the service quality of geographically separated UE and enhances the overall system performance [12].

Cognitive Radio: Over the spectrum has been allocated for several usages and the allocated resources are very much under-utilized, To reduce spectrum scarcity and utilize the underutilized spectrum, cognitive radio technology has been proposed. It is an intelligent radio, that can be sense, learn, aware and adapt according to the environment. With the help of software-defined radio (SDR), cognitive radio can be programmed and configured dynamically. SDR is a radio transceiver where radio components (modulators/demodulators, filters, amplifiers, mixers, detectors, etc.) are implemented by software on a personal computer or embedded system [12].

Having understood these technologies of 5G wireless communication, FCN is planning to have the communication system that can achieve data rates of about 100 Tb/s high speed, low latency, and reliable communications are essential for supporting ML/AI at the edge; giving rise to the research field entitled Communication over machine learning. Incorporation of holographic telepresence, holographic communication, virtual reality, and augmented reality in future communication, boost the requirement of wireless communication [14].

2.2 Development of optical network

In the present network scenario, the data generated by the wireless devices are transported through an optical network. In general, optical fiber is connected between wireless base stations (BSs/eNB), and their controlling, switching, and monitoring centers. Due to the enormous available bandwidth, the optical fiber can carry data up to 100 Tbps for networking in the optical network. By using appropriate technology, the capacity can be increased further. Similar to the evolution of the wireless network, the optical networks also evolved generation-wise. During the process of evolution, the optical network incorporated optical cross-connect (OXC), a synchronous digital hierarchy (SDH) /synchronous optical network (SONET) rings, optical add-drop multiplexers (OADMs), Software-defined network/network function virtualization (SDN/NFV). Today's long-haul backbone networks of 10/40 Gbps wavelength channels use wavelength-division multiplexing (WDM) transmission systems. Further increase in capacity, the optical network uses a dense WDM (DWDM) frequency grid (12.5, 25, 50, and 100 GHz by G.694.1). Further development of WDM transmission systems makes the system an adaptable DWDM grid.

Optical Transport Network (OTN): ITU-T G.709 defined OTN, which transport digital/optical signal across the core network is a flexible way. Each optical channel carries a separate signal using optical channels multiplexing and uses optical data as a unit. OTN supports the different functions for transporting data, such as multiplexing, routing, management, supervision, and survivability.

Automatically Switched Optical Network (ASON): To accommodate dynamic traffic and their requirements, optical networks need to manage to signal and routing automatically and intelligently. It provides auto-discovery and dynamic connection set-up with the help of dynamic signaling-based over OTN and SDH networks. This is done through a distributed (or partially distributed) control plane, which enables improved support for current end-to-end provisioning, re-routing, and restoration. ASON uses the generalized MPLS (GMPLS) signaling protocol to set up and monitor edge-to-edge transport connections. It also uses single fiber switching to wavelength switching and optical packet switching. The other components, like OXCs, wavelength converters, and OADMs are required for ASON.

Different variant of wavelength-division multiplexing (WDM): WDM is the main transmission technology. Over the year several of its variant has been proposed and used, which are Dense WDM, Coarse WDM, and Time WDM. DWDM uses frequency grids of 12.5, 25, 50, and 100 GHz for transmission. IN the present scenario, many-core networks deployed 1.6 Tbps ($40\,\text{Gbps} \times 40$ wavelengths) DWDM system. To support the capabilities for 5G and beyond 5G system, the core network will need to transport 10 Tbps or more per fiber which will be pushed further for future FCN.

CWDM combines multiple optical signals at various wavelengths for transmission in optical fiber cables. Up to 18 channels are allowed to be connected over a dark fiber pair. Unlike 0.4 nm spacing for DWDM, CWDM systems have channels at wavelengths spaced 20 nanometers (nm) apart. CWDM works well in two prominent wavelength regions, 1310 nm, and 1550 nm.

TWDM is a WDM technique, where TDMA is applied to a set of wavelengths instead of just one wavelength. It requires strict coordination with the radio equipment to guarantee low latency, as with TDMA and provides more bandwidth than TDMA. In a passive optical network (PON), TWDM can be used as an alternative for transmitting 5G traffic [14].

Enhanced Common Public Radio Interface (eCPRI) fronthaul: CPRI is the key internal interface of Radio Equipment (RE), or remote radio head (RRH) and base station unit (BBU) or radio equipment controller (REC) via fronthaul transport network. For fronthaul between RRH and BBU, the overall delay must be limited to less than 100 μs over the multi-hop paths in 5G communication. Due to this stringent latency requirement, eCPRI is becoming an important technology for 5G. Its specification supports more flexibility in the positioning in eNBs, where BBU contains part of the PHY layer and higher layer functions of the air interface, whereas the RRH contains the remaining part of the PHY layer functions and the analog radio frequency functions [15].

Software-Defined Optical Network: The SDN paradigm separates the control plane from the data plane and uses an SDN controller for centralizes network control. SDN facilitates NFV for the network virtualization over the physical infrastructure so that multiple virtual networks can operate within. Due to high optical transmission capacities and the specific characteristics of optical components, software-defined optical networks (SDONs) has been proposed. With an underlying optical network infrastructure, SDONs seek to leverage the flexibility of SDN control for supporting networking applications. NFV allows for the flexible operation of multiple virtual optical networks over a given physical optical network infrastructure. SDONs are highly promising for low-latency and high-bandwidth backhauling for 5G eNBs. SDON application layer studies have developed mechanisms for achieving Quality of Service (QoS), access control and security, as well as energy efficiency and failure recovery [16].

Reconfigurable Optical Add/Drop Multiplexers (ROADM): OADM drops the desired wavelengths to local terminals from an incoming multi-wavelength signal by using a wavelength demultiplexer and adds a locally generated wavelength with the remaining pass-through wavelengths to generate the new outgoing multi-wavelength signal. In general, the mux/demux characteristics are fixed. However, to accommodate dynamic behavior and requirements of an optical network, it is almost necessary to have a reconfigurable OADM (ROADM). A ROADM can switch traffic remotely from a WDM system at the wavelength layer and enables the flexibility and reconfigurability of an optical transport network. Having the properties of being colorless (not wavelength selective), directionless (not nodal degree selective), and contentionless (not different wavelength) improves significantly the capacity of add/drop ports in a ROADM [17].

Software-Defined Optics (SDO): Due to dynamic and variable requirements of data traffic, it is almost necessary to do cross-layer interactions. To enables this

SDO has been in an optical network. This can be done through the construction of application-specific protocol stacks out of small reusable services [1].

Elastic optical network (EON): EON accommodates dynamic changes of the optical component, such as flexible wavelength assignment, redefined optical switches and various transponders, etc. to improve system performances. It is possible to integrate EON with an IP layer easily to construct an IP-over-EON. The basic unit of switching in EON is a sub-carrier instead of the wavelength in the fixed-grid case since channels are usually composed of a variable number of sub-carriers [18].

Software-defined optical transmission (SDOT): SDOT supports dynamic reconfigurability of optical components and the ability to adapt various transponders. Softwarization and intelligent control of the data plane facilitates SDOT to optimal use of the available resources, which exploits the multiple dimensions and granularities efficiently.

Cognitive optical network (CON): To make more agile optical networking, a cognitive optical network has been proposed. The CON architecture enables high data rate lightpaths while compensating for a variety of dispersion impairments. CON improves the measuring parameters [e.g., optical signal-to-noise ratio (OSNR), chromatic dispersion (CD), polarization-mode dispersion (PMD), and bit error rate (BER)] while compensates the impairment and subsequently enhance the QoT in the optical network efficiently [19].

3. Performance requirement for end-to-end services

Nowadays, end-to-end performance is based on customer experience. As the data is sent over a heterogeneous network combination of wireless and wired (as shown in **Figure 4**), passing through several autonomous systems. These are operated by the same or different operators, which are using various networking technologies. These connections are inter-technical, inter-national, and

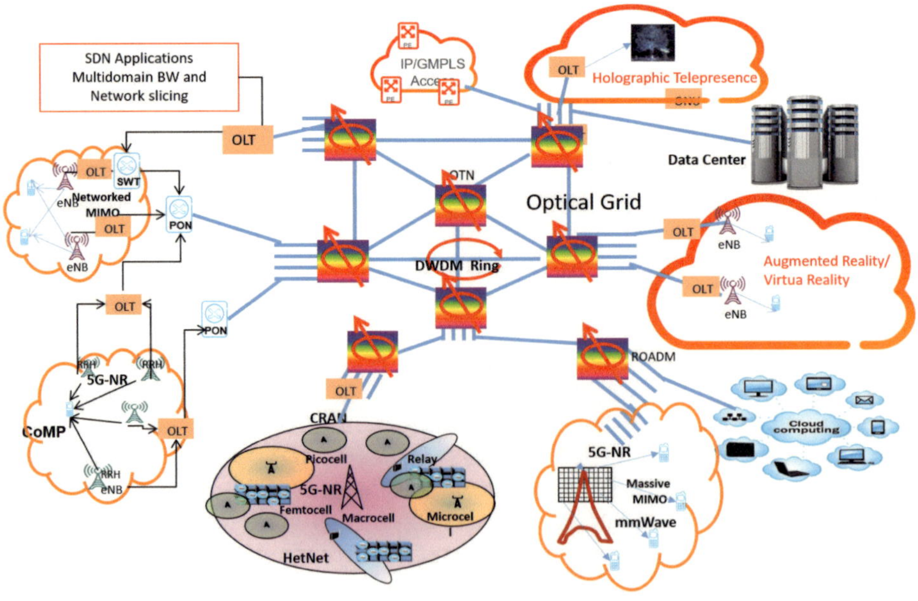

Figure 4.
A typical service architecture of a wireless access network with an optical backbone.

inter-continental. Thus end-to-end performance is measured by the quality of experience (QoE) along with other metrics like quality of service (QoS), quality of resilience (QoR), quality of transmission (QoT), etc.

QoS/QoE parameters are different for different applications. Varies in latency, connectivity, data rate, etc. QoE evaluation by the user depends on several independent factors, such as service type, user profile (details of user personal information), type of equipment, type of content or service pricing policy (free, paid), screen size, etc. QoE is influenced not only by QoS but also by the grade of service (GoS) and QoR, as shown in **Figure 5**. The most popular measure of QoE is based on the Mean Opinion Score (MOS) [20].

User experience varied QoE while using the service from a different operator. At the technology level, operators are launching new services, which can work with virtualized, software-based, cloud-native, and more agile networks. In general, customer's QoS/QoE needs to be monitored across physical. In virtualized networks, this becomes even more critical where services will be activated in real-time and need to be tested, fulfilled, and assured in an automated fashion.

The specification of 5G communication is different for different applications, like M2M, high broadband, and uRLLC. All these applications have different requirements (see **Figure 2**). Apart from these, in 6G communication, several new applications have been proposed, such as holographic telepresence, AR/VR, etc. The transmission requirements of these applications are quite futuristic in terms of data rate, latency, and BER. Thus, maintaining QoE in FCN will be very complex and challenging, as there are different types of CoS asking for separate GoS working in various environments, policies, and networks.

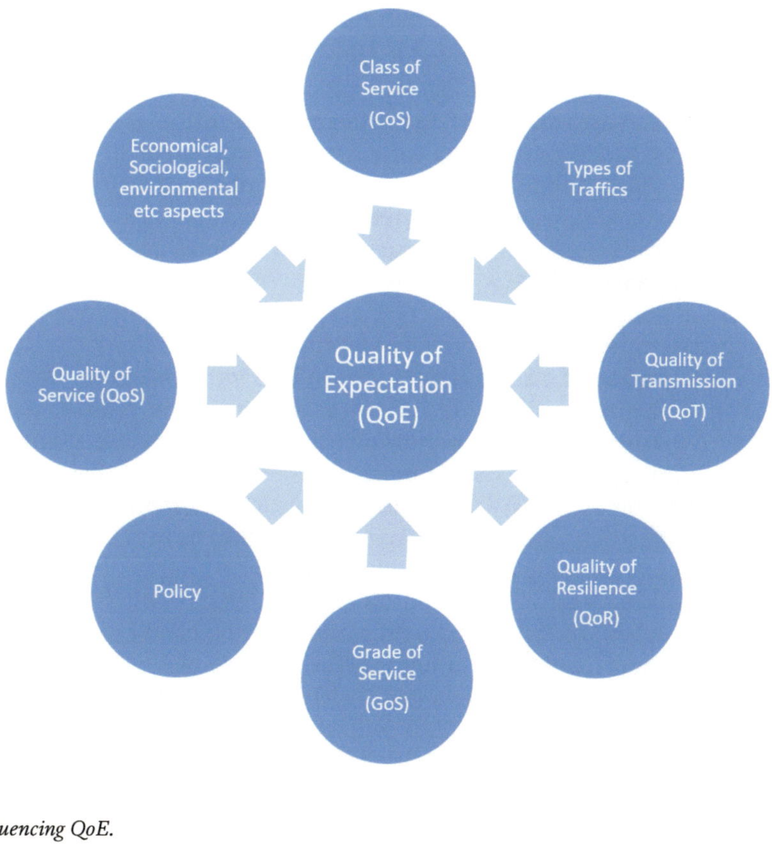

Figure 5.
Factors influencing QoE.

4. Behavior of evolved wireless technologies with corresponding evolved optical techniques to satisfy user QoE

In the present day, users require appropriate supports from the network infrastructure as per the service usages. In general, users are connected to the network through wireless access and the wireless access point is connected to the optical fronthaul node. Depending on the application, the user required variable BW, the latency of wireless access to satisfy its QoE. Optical network technologies will play an important role in addressing these requirements within the radio access network (RAN). Through the deployed network technologies, such as backhaul networks, metro networks, and PONs, etc., optical networks continuously support their QoS. The optical network used an eCPRI fronthaul interface to support the 5G specification. For example, eCPRI of 100 Gb/s supports a 5G system of 200 MHz BW (below 6 GHz frequency) with 64 (8X8) antenna arrays. It can also support mmWave communication of 400 MHz radio bandwidth in 60 GHz frequency range with 256 (16x16) radiating elements by 400–800 Gb/s capacity [21, 22].

Support IoT Applications: I0T is one of the widely used technology in recent times, which is used in a wide variety of applications. The use of IoT appears to be most challenging due to the wide range of different devices, various options of network connectivity, different protocols, methods, etc. It provides support to users with smart services while raising security and privacy threats [23]. The threat becomes challenging while users and networks are heterogeneous. To support this heterogeneity in IoT, SDO provides an appropriate solution. A solution like cognitive radio and CON can work together to facilitate the dynamic behavior and requirements of diverse IoT applications. Apart from this, SDN, wireless-SDN and SDON are also participating to support IoT services, while using edge router to integrate into the network.

Reduce outage of edge users: To reduce the outage of edge users, successful operation of CoMP is necessary, which depends on very fast and highly reliable feedback between the user and eNBs on the channel condition. At the same time, all the eNBs need to be synchronized and data should be present at all eNBs in real-time. Connecting optical fiber link between eNBs should ensure this low latency level as per 5G standard through the fast feedback channel. This is more complex, challenging when the number of participating eNBs is more, and the traffic load of the network increases. These will impact on processing (impact on delay in data transmission), synchronization (impact on a real-time mismatch), which depends on deploying sites topology, backhaul latency, and capacity [23].

Flexible Integration of data traffic: In 5G and 6G wireless communication, the data traffic has a diverse specification and has a wide variety of requirements. To support these dynamic and diverse requirements, flexibility and adaptability should be supported by an optical network. Optical networks supported these flexible and elastic nature by using SDN/NFV. Integration of optical components, such as various variants of ROADMs, OXC in multi-layer SDN makes network towards SDON. The use of different switching paradigms and a combined implementation of the switching elements in electronics and optics (hybrid optical switching) in SODN, can lead to even higher flexibility and better transmission efficiency [24].

Integration to heterogeneity: To support the requirement of 5G, the standards like NG-EPON (by IEEE) and G.hsp.x (by ITU-T) are proposed. Coexistence of 10G PON channels for residential, 100G dedicated channel for business along with wireless fronthaul, supports heterogeneity of 5G & beyond 5G communication. These are supported by long reach TDM-DWDM PON system, with up to 100 km reach, 512 users, and an emulated system load of 40 channels, employing amplifier nodes with either erbium-doped fiber amplifiers (EDFAs), or Raman amplifier or semiconductor optical amplifiers (SOAs). This end-to-end support by SODN with help of PON physical layer along with dynamic wavelength allocation (DWA) in response to increased traffic demand [25].

Service on the fly: Providing service through the cloud is immensely popular among users. The 5G communications also advocate and support the application through the cloud and aim to provide them effectively and efficiently. The Cloud-RAN (C-RAN) approach for 5G wireless splits the radio processing chain to simplify the processing. To optimized support for different technologies, levels of centralization, and deployment options in 5G, EONs offer large degrees of flexibility, adaptability, and programmability in different dimensions. EON provides granular spectrum width consisting of variable numbers of sub-carriers as the demand and deployment technology to support 5G disruptive capabilities, technologies, and use cases. It allows both digital and analog signals to be transported and switched over the same optical fiber, thus facilitating technologies such as mm-wave. Besides, the EON can tune signal properties (e.g., modulation format, bit rate, optical reach, and so on) to cope with the constraints of deployed technologies and different requirements of use cases [26, 27]. Hence, it can provide a much larger bandwidth and more variety of bit rates on an optical fiber.

Enabling Artificial intelligence (AI): In the present times, AI has taken center stage in all kinds of research and development. To provide better support, monitor, and control, every kind of service uses AI technology. AI will learn with the help of a machine learning algorithm for better service. AI and machine learning is the main technology of 5G and 6G communication. Wireless and optical networks use these technologies extensively. Specifications like high speed, low latency, and reliable communications are essential for supporting ML/AI at the edge. This can bring mobile edge computing to AI-at-the-edge [28, 29].

Energy Consumption: As the requirements are increasing to satisfy enhance throughput, latency & other QoS for different classes of traffic, applications, services, and QoE, energy consumption is increasing in the network. The usage of massive MIMO, dense network, heterogeneous network with small cells along with billions of devices increases the power consumption in the network, which increases the greenhouse effect. 5G power consumption at peak hours is 1200 W to 1400 W, which is 300–350% greater than of 4G [30]. However, to work in an energy-efficient way, network wire-line, wireless and core networks) are using the resources in an optimized way, which motivates the network to use different tradeoffs in protocol layers [31], which varies from infrastructure (dense network) to device (visual resolution). A green framework has been proposed for energy-efficient communication in a wireless network with the energy-cognitive cycle, where the awareness is categorized as network awareness and access point awareness module [32].

Every component of the optical network participated in the data transports and consumes energy. The CAPEX amount is more at the beginning, however, usage of massive MIMO along with small cells in dense networks can impact more on OPEX. Usage of NFV decreases the OPEX costs by reducing the conventional purposed hardware, installation, and up-grading for new services and Virtual network functions (VNF) are virtualized tasks implemented by the NFV platform, providing security, load balancing, and other EPC functions [33]. A WDM transmitter/receiver (TX/RX) pair at the interface between each link provides regenerated signals at each wavelength for injection in the next link of the system. Energy consumption exists at many levels in optical transmission systems, from inefficiencies at the device level in optical amplifier pump lasers and their cooling systems, at the circuit level in the tradeoff of efficiency for speed in high-speed electronic circuits used in transmitters and receivers, and at the system level in terms of multiplexing and management overheads [34, 35].

5. Conclusion

This chapter provides an overview of telecommunication networks while considering both wireless and optical networks. The technologies in both the networks were evolved in such that they can assist each other for better performance of users and network as a whole. The chapter provides an overview of the main technologies in 5G and analyses how the optical network technologies are beneficial and cooperative to wireless technologies.

Author details

Rajarshi Mahapatra
Dr. SPM International Institute of Information Technology, Naya Raipur, India

*Address all correspondence to: rajarshim@ieee.org

References

[1] Nina Slamnik-Krijestorac, Haris Kremo, Marco Ruffini, Johann M. Marquez-Barja, Sharing Distributed and Heterogeneous Resources toward End-to-End 5G Networks: A Comprehensive Survey and a Taxonomy, IEEE Communications Surveys & Tutorials, volume: 22, Issue: 3, pp. 1592-1628 third quarter 2020),

[2] Yangyishi Zhang; Rong Zhang; Jiankang Zhang et al, Far-End Crosstalk Mitigation for Future Wireline Networks Beyond G.mgfast: A Survey and an Outlook, IEEE Access (Volume: 8), January 2020

[3] Roberto Sabella, Paola Iovanna, Giulio Bottari, Fabio Cavaliere. "Optical transport for Industry 4.0", Journal of Optical Communications and Networking, Vol. 12, No. 8, pp. 264-276, August 2020,

[4] Ioannis Tomkos, Dimitrios Klonidis, Evangelos Pikasis, Sergios Theodoridis. "Toward the 6G Network Era: Opportunities and Challenges", IT Professional, 2020

[5] Rajarshi Mahapatra. "Participation of Optical Backbone Network in Successful Advancement of Wireless Network", Wireless Personal Communications, 2017

[6] Shao-Yu Lien, Shao-Chou Hung; Der-Jiunn Deng; Yueh Jir Wang, Efficient Ultra-Reliable and Low Latency Communications and Massive Machine-Type Communications in 5G New Radio, GLOBECOM 2017-2017 IEEE Global Communications Conference, 4-8 Dec. 2017

[7] M. A. Siddiqi, H. Yu and J. Joung, "5G ultra-reliable low-latency communication implementation challenges and operational issues with IoT devices", Electronics, vol. 8, no. 9, pp. 981, Sep. 2019.

[8] Ali Zaidi, Fredrik Athley, Jonas Medbo, Ulf Gustavsson, Giuseppe Durisi, Xiaoming Chen. "Multiantenna Techniques", Elsevier BV, 2018

[9] Jungnickel, V., et al. (2014). The role of small cells, coordinated multipoint, and massive MIMO in 5G. IEEE Communications Magazine, 52(5), 44-51. 2014

[10] Dai, L., Wang, B., Yuan, Y., Han, S., Chin-Lin, I., & Wang, Z. (2015). Non-orthogonal multiple access for 5G: Solutions, challenges, opportunities, and future research trends. IEEE Communications Magazine, 53(9), 74-81.

[11] M. Agiwal, A., Roy, & Saxena, N. (2016). Next generation 5G wireless networks: A comprehensive survey. IEEE Communications Surveys & Tutorials, 18(3), 1617-1655.

[12] Chand, P., Mahapatra, R., & Prakash, R. (2016). Energy efficient radio resource management for heterogeneous wireless network using CoMP. Wireless Network, 22(4), 1093-1106

[13] G. A. Akpakwu et al.:, A Survey on 5G Networks for the Internet of Things: Communication Technologies and Challenges, IEEE Access, vol 6, pp. 3619-3647, Jan 2018.

[14] Paola Iovanna, Fabio Cavaliere, Stefano Stracca, Luca Giorgi, Fabio Ubaldi. "5G Xhaul and Service Convergence: Transmission, Switching and Automation Enabling Technologies", Journal of Lightwave Technology, 2020

[15] S. Bjørnstad, R. Veisllari, D. Chen, F. Tonini, C.Raffaelli. "Minimizing Delay and Packet Delay Variation in Switched 5G Transport Networks", Journal of Optical Communications and Networking, 2019

[16] Akhilesh S. Thyagaturu, Anu Mercian, Michael P. McGarry, Martin Reisslein, Wolfgang Kellerer. "Software Defined Optical Networks (SDONs): A Comprehensive Survey", IEEE Communications Surveys & Tutorials, 2016

[17] Yongcheng Li, Jingjing Li, Liangjia Zong, Sanjay K. Bose, Gangxiang Shen. "Upgrading Nodes with Colorless, Directionless, and/or Contentionless ROADMs in an Optical Transport Network", 2020 22nd International Conference on Transparent Optical Networks (ICTON), 2020

[18] Raouf Boutaba, Nashid Shahriar, Siavash Fathi. "Elastic Optical Networking for 5G Transport", Journal of Network and Systems Management, 2017

[19] Wei, W., Wang, C., & Yu, J. (2012). Cognitive optical networks: Key drivers, enabling techniques and adaptive bandwidth services. IEEE Communication Magazine, 50(1), 106-113

[20] Stankiewicz, R., Cholda, P., & Jajszczyk, A. (2011). QoX: What is it really? IEEE Communications Magazine, 49(4), 148-158.

[21] Paola Iovanna, Fabio Cavaliere, Stefano Stracca, Luca Giorgi, Fabio Ubaldi. "5G Xhaul and service convergence: transmission, switching and automation enabling technologies", Journal of Lightwave Technology, 2020

[22] Gabriel Otero Perez, David Larrabeiti Lopez, Jose Alberto Hernandez. "5G New Radio Fronthaul Network Design for eCPRI-IEEE 802.1CM and Extreme Latency Percentiles", IEEE Access, 2019

[23] Shancang Li, Li Da Xu, Shanshan Zhao. "5G Internet of Things: A survey", Journal of Industrial Information Integration, 2018

[24] Francesco Musumeci, Omran Ayoub, Monica Magoni, Massimo Tornatore. "Latency-Aware CU Placement/Handover in Dynamic WDM Access-Aggregation Networks", Journal of Optical Communications and Networking, 2019

[25] Shijian Gao, Xiang Cheng, Liuqing Yang. "Estimating Doubly-Selective Channels for Hybrid mmWave Massive MIMO Systems: A Doubly-Sparse Approach", IEEE Transactions on Wireless Communications, 2020

[26] Dawit Hadush Hailu, Berihu G. Gebrehaweria, Samrawit H. Kebede, Gebrehiwot G. Lema, Gebremichael T. Tesfamariam. "Mobile fronthaul transport options in C-RAN and emerging research directions: A comprehensive study", Optical Switching and Networking, 2018

[27] Isiaka Ajewale Alimi, Antonio Luis Teixeira, Paulo Pereira Monteiro. "Toward an Efficient CRAN Optical Fronthaul for the Future Networks: A Tutorial on Technologies, Requirements, Challenges, and Solutions", IEEE Communications Surveys & Tutorials, 2018

[28] A. Z Azzaouri et al, "Block5GIntell: Blockchain for AI-Enabled 5G Networks," IEEE Access, vol 8, August 2020

[29] Faris B. Mismar et al, Deep Reinforcement Learning for 5G Networks: Joint Beamforming, Power Control, and Interference Coordination, IEEE Transaction on communication, vol 68, no. 3, March 2020

[30] A. Mughees et.al, Towards Energy Efficient 5G Networks Using Machine Learning: Taxonomy, Research Challenges, and Future Research Directions, IEEE Access, vol8, pp. 187498-187522, Oct 2020

[31] Mahapatra, R., et al. (2016). Energy efficiency tradeoff mechanism towards

wireless green communication: A survey. IEEE Communication Survey and Tutorial, 18(1), 686-705. (First Quarter).

[32] Mahapatra, R., et al. (2013). Green framework of future heterogeneous wireless network. Computer Network, 57(6), 1518-1528.

[33] A. N. Al-Quzweeni, A. Q. Lawey, T. E. Elgorashi, and J. M. Elmirghani, "Optimized energy aware 5G network function virtualization," IEEE Access, vol. 7, pp. 4493944958, 2019.

[34] Tucker, R. S. (2011). Green optical communications. Part I: Energy limitations in transport. Part II: Energy limitations in network. IEEE Journal of Selected Topics in Quantum Electronics, 17(2), 245-274.

[35] Tucker, R. S. (2011). Green optical communications. Part II: Energy limitations in network. IEEE Journal of Selected Topics in Quantum Electronics, 17(2), 245-260.

Chapter 6

Near-Field Communications (NFC) for Wireless Power Transfer (WPT): An Overview

Poonam Lathiya and Jing Wang

Abstract

Recent advancements in the semiconductor integrated circuits and functional materials technologies have accelerated the demand of electronic and biomedical devices such as internet of things (IoT) and wearable sensors, which have low power consumption, miniature size and high data transfer efficiency. Wireless power transfer (WPT) has become the alternative solution to current electronic devices that rely on bulky batteries to supply the power and energy. Near Field Communication (NFC) technology is extensively used for wireless power transfer, where devices communicate through inductive coupling via induced magnetic fields between transmit and receive coils (loop antennas). Thin NFC sheets made of soft magnetic materials are inserted between antennas and metal case of wireless gadgets, such as mobile phones or tablets, to reduce the degradation of antenna gain and radiation efficiency due to generation of eddy currents. To enhance the efficiency of wireless power transfer, magnetic materials with superb properties such as high permeability, low magnetic loss and high resistivity are highly desirable. In this chapter, we will provide an overview of the current state of the art, recent progress and future directions in NFC based wireless power transfer, with the special focus on near field communications operating at 13.56 MHz.

Keywords: wireless power transfer, near field communication, inductive coupling, ferrites, power transfer efficiency

1. Introduction

At the start of 21st century, an emerging technology known as near field communication (NFC) was standardized in 2004 that gradually changed the consumer electronics market and facilitated the electronic transactions, mobile payments, data transfer, etc. NFC was first reported by joint venture of Sony and NXP Semiconductors in 2002 [1]. Since then, NFC has become a popular and evolving technology over the past decade and is being incorporated into more and more aspects of our daily lives than ever before. The NFC Forum was founded in 2004 by joint venture of Sony, Philip and Nokia to facilitate the enhancement of this NFC technology. Nokia 6131 was the first NFC enabled device which was launched in 2006 [2, 3]. In 2006, NFC technology was utilized to print disabled patrons in the libraries with enabled disables patrons [3]. In 2011, University of Bristol's M-Biblio started NFC enabled QR codes for students to utilize library resources. Samsung launched first NFC enabled android (Samsung NEXUS S) phone in 2010 [2].

Area	Growth (2018–2019)
NFC interactions	27%
NFC activators	22%
NFC reach	50%
No. of interactions per active NFC object	6%

Table 1.
NFC usage growth evolution from 2018 to 2019 [6].

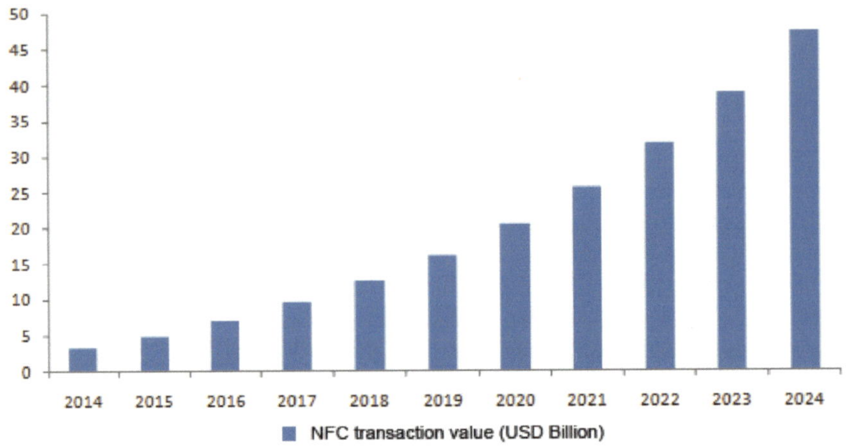

Figure 1.
Evolution of NFC transactions values between years 2014 and 2024 [7].

In 2011, PayPass functionality was launched for RIM's (Research in Motion/BlackBerry Limited) master card. Some of the early applications launched are, Samsung TecTile Programmable NFC tags in 2012, Sony's Xperia smart tags, NFC enable Smart Objects in 2011, NFriendConnector [4], Wallet in 2011, the joint venture of Google At&T, Verizon and T-mobile in 2012 to use mobile wallets [5]. NFC enabled functionality has been added to all new Apple products starting from iPhone XS (Apple Pay). **Table 1** shows NFC usage growth evolution in 2018–2019 [6]. From 2010 onward, new interesting applications of NFC was launched every year in communication sector by technology giants such as Google, Apple, Samsung, NXP, etc. The industry players are constantly introducing new advances and improved technologies in NFC enabled devices which have taken global market to 4.80 billion USD in 2015 and expected to reach 47.42 USD billion by 2024. **Figure 1** shows projected NFC transactions value from 2014 to 2024 [7].

2. Near field communication magnetism – NFC and RFID

NFC enables a subset of the Radio Frequency Identification (RFID) technology that works over a wide range of frequencies with three distinct bands — low, high, and ultra-high frequencies. The main difference between NFC and RFID technologies is their operating range. RFID operates in meters range, whereas NFC typically operates within three to five centimeters. All RFID's operate based on the same principle of one-way data transfer from the tag to the receiver and there is no power transfer the other way around [8, 9]. RFID is one of the oldest technologies that

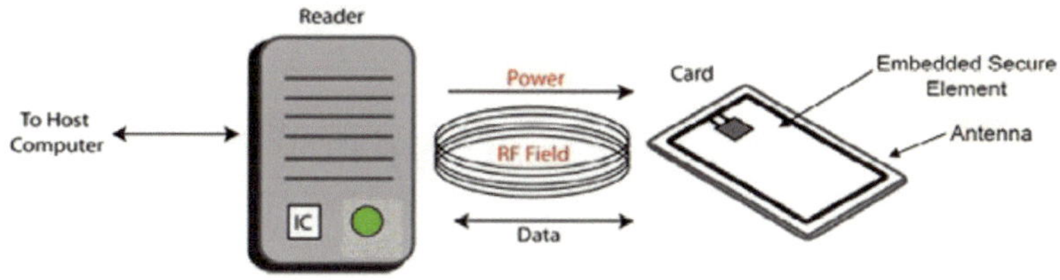

Figure 2.
Contactless transfer of data/signal between reader and tag at 13.56 MHz using NFC technology [16].

utilize near field magnetic communication. In 1960, an electronic articles surveillance system (EAS) was the first commercial application of RFID, which utilized one-bit tag and was used to detect the presence or absence of the tag. Between 1970 to 1980, more work on RFID systems was conducted utilizing microwave and inductive systems and, in late 1970s, the size reduction of RFID's was accomplished using low-power complementary metal-oxide semiconductors (CMOS) logic circuits. After 1980, RFID applications became widespread such as tracking for animals, business, electronic toll collection, and automation, which was rapidly expanded with the development of personal computer (PC) technology. In 1990s, electronic toll collection systems were the first successful application of RFID technology worldwide [10]. Presently, RFID is utilized in various commercial areas such as automobile, agriculture, transport, medical system, payment cards, supply chain, tracking, identification application and short range interactions in the Internet of Things (IoT) [11, 12]. However, communications which require initialization at both ends (e.g., Peer-to-peer communications as discussed below) cannot be supported by RFID technology. NFC is a great solution to this shortcoming of RFID, which support peer-to-peer communications also.

NFC is a short-range half duplex communication technology which provide secure communication between devices in near field region. Near field communication, is a technology that allows two devices in close range to securely exchange data wirelessly. NFC is a short-range (< 10 cm) wireless connectivity technology that operates at high frequency (HF) range with low bandwidth of radio waves, mainly at 13.56 MHz [13]. NFC comprises of three basic components - an antenna, a reader, and a tag. A reader (transmitter) sends a signal at the standard NFC frequency of 13.56 MHz and the tag antenna receives and processes the interrogation signal, and responds with requested information back to the reader that is then interpreted and stored as the data within few centimeters at 13.56 MHz [14, 15]. **Figure 2** shows the transfer of data between reader and tag (card) at 13.56 MHz based on NFC technology [16].

From 2004 onwards, NFC has been utilized in various applications. Nokia, Apple/Google/Samsung pay transactions, wireless energy/data transmission and wireless key card entry are a few popular examples of this technology [17]. Though NFC tag is passive in nature, NFC can transfer data both ways. NFC technology supports varying data transmission rates, typical three rates are - 106, 212 and 424 Kbps [18]. There is another 848 Kbps rate also, but it is not in compliant with NFC standards.

3. Basic principle of NFC

NFC works based on the principle of near field magnetic communication. This principle of inductive coupling is applied to all communications based on near field magnetism between transmitting and receiving devices. **Figure 3** shows the

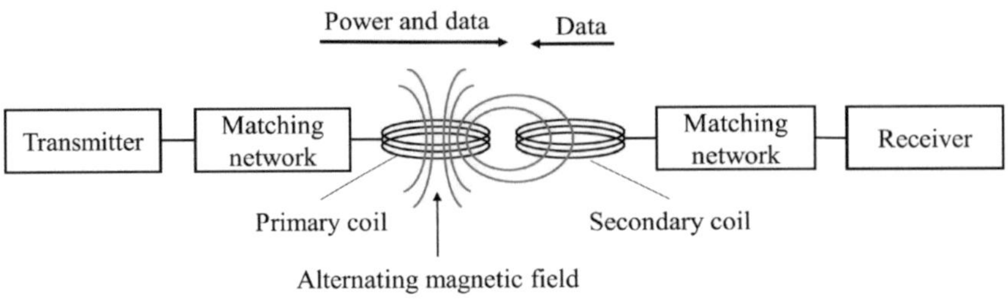

Figure 3.
Inductive coupling between transmitter and receiver coils [9].

simplified concept of inductive coupling. When a primary coil generates alternating magnetic field, secondary coil which is placed in the vicinity of the primary coil inductively coupled with the primary coil and generates induced alternating magnetic field according to the Faraday's law. This is the basic principle in transferring power wirelessly between the devices in near field region. The above stated principle also applied on RFID systems that are based on inductive coupling. Even though, there are some differences in other components such as network system and protocols between NFC and general RFID systems.

3.1 Inductive coiled system

The inductive coupled NFC system can be modeled by using expressions of self-inductance, mutual inductance and resistances [19]. A generalized analytical expression for calculation of self-inductance of circular or rectangular shaped coil is explained below. **Figure 4** shows a representation of single turn circular coil while illustrating the magnetic field pattern surrounding two circular coils [20].

The inductance, L_0, for a single turn circular coil can be given by Eq. (1) as seen below [21]:

$$L_0 = \mu_0 r ln\left(\frac{2r}{d}\right) \tag{1}$$

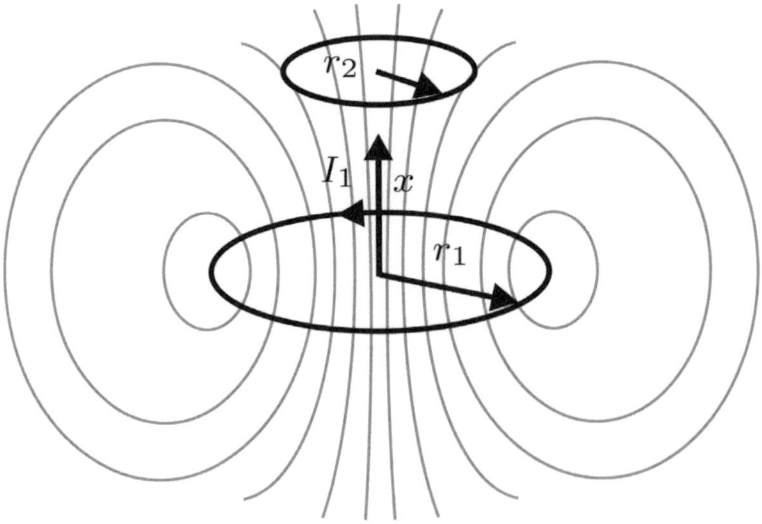

Figure 4.
Depiction of two magnetic inductive coil system (primary and secondary) [20].

Where μ_0 is permeability of free space, r is radius of the coil and d is the diameter of the wire. The single turn coil inductance can be utilized to calculate inductance of multiturn coil, L and is given by Eq. (2).

$$L = N^2 L_0 \tag{2}$$

$$L = N^2 \mu_0 r ln\left(\frac{2r}{d}\right) \tag{3}$$

where N is number of turns in a coil. This equation provides appropriate approximation for a cylindrical inductor, but for the case of spiral inductor this equation provides general parameter studies. A detailed study for calculation of inductance of spiral inductors is provided by S. Alturi et al. (2004) [22].

Another important figure of merit is mutual inductance of two coupled coils. The mutual inductance between the two coils can be expressed in Eq. (4).

$$M = \frac{\mu \pi N_1 N_2 r_1^2 r_2^2}{2\sqrt{\left(r_1^2 + x^2\right)^3}} \tag{4}$$

where, N_1 and N_2 are the number of turns in first and second coil, respectively, r_1 and r_2 are the radius of first and second coil, respectively, x is the axial separation and μ is the permeability. Eq. (4) is valid only for $r_2 < r_1 \ll x$, i.e., the magnetic field generated by current I_1 in first coil should be homogeneous in the mutually bounded area of the 2nd coil. Detailed calculation for mutual inductance for cylindrical coils is presented in [23].

Also, the coupling factor between 2 coils can be expressed in Eq. (5).

$$k = \frac{M}{\sqrt{L_1 L_2}} \tag{5}$$

Where k is coupling factor and lies between 0 and 1, L_1 and L_2 are inductance of first and second coil, respectively.

In case of complex geometries, numerical methods can be applied to calculate inductances of complex coil systems [24].

3.2 Wireless power transfer (WPT) efficiency

Power transfer efficiency between loop antennas for NFC system is expressed as the figure of merit that depends on inductive coupling. As NFC operates at small distance range between transmitter and receiver antennas, its efficiency depends on coupling between the antennas for wireless power transfer [25]. **Figure 5** shows schematic of two mutual magnetic coupled coil antennas for the WPT system [27]. To improve the power efficiency, an impedance matching on both coil antennas (receiver and transmitter) is required. In case of magnetic coupling between receiver and transmitter coil antennas, eddy currents are generated due to alternating magnetic field. This cause a shift of resonant peaks of the input impedance, which shifts resonance frequency of maximum power transfer. Insertion of a high permeability soft magnetic ferrite sheet between coil antenna and metal conductor shifts the frequency back to original resonance frequency [26]. When both circuits resonate at peak frequency, maximum transfer of power is achieved. **Figure 6** depicts the simplified equivalent circuit model of wireless power transfer systems [26].

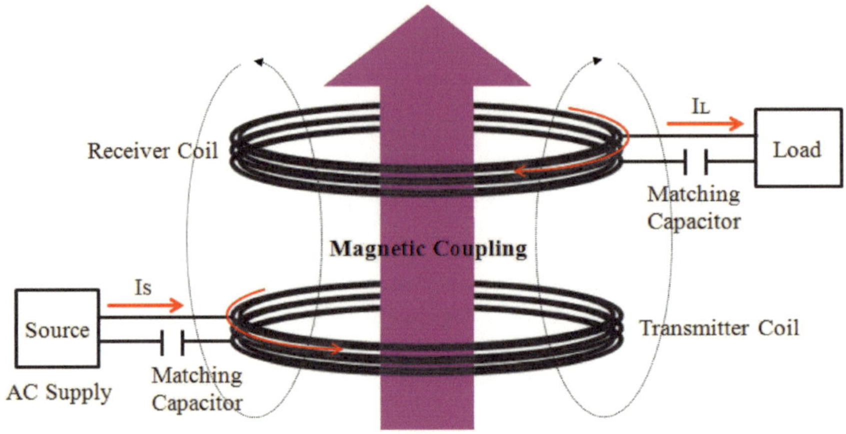

Figure 5.
Schematic drawing of mutual magnetic coupled coils for wireless power transfer systems [26].

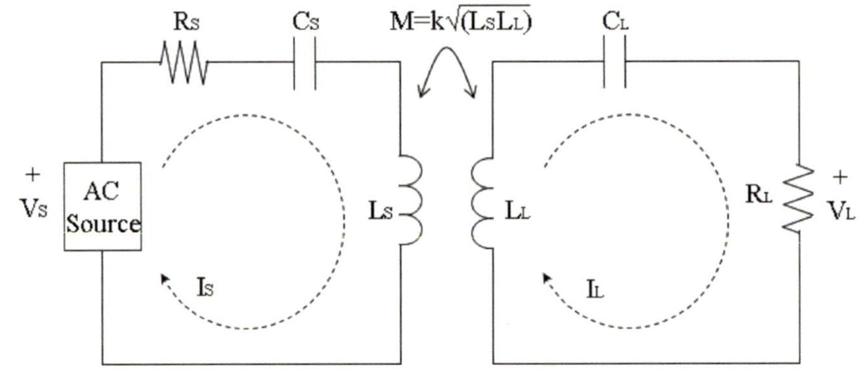

Figure 6.
A simplified equivalent circuit model of wireless power transfer systems [27].

The resonance frequency ω for coil antenna is given by,

$$\omega = \frac{1}{\sqrt{L_s C_s}} = \frac{1}{\sqrt{L_L C_L}} \tag{6}$$

Base on Kirchhoff voltage law and above circuit diagram, V_s can be calculated as [28]:

$$\begin{bmatrix} j\omega L_s + \dfrac{1}{jwC_s} + R_s & j\omega M \\ j\omega M & j\omega L_L + \dfrac{1}{j\omega C_L} + R_L \end{bmatrix} \begin{bmatrix} I_s \\ I_L \end{bmatrix} = \begin{bmatrix} V_s \\ 0 \end{bmatrix} \tag{7}$$

From Eq. (7), V_s can be calculated [29]:

$$V_s = I_s\left(R_s + j\omega L_s + \left(\frac{1}{j\omega C_s}\right) - I_L(j\omega M)\right. \tag{8}$$

$$0 = I_L\left(j\omega L_L + \left(\frac{1}{j\omega C_L}\right) + R_L\right) - I_s(j\omega M) \tag{9}$$

Using Eq. (8) and (9):

$$I_L = I_s\left(\frac{j\omega M}{j\omega L_L + \left(\frac{1}{j\omega C_L} + R_L\right)}\right) \tag{10}$$

Substituting Eq. (10) into Eq. (8):

$$V_s = I_s\left(R_s + j\omega L_s + \left(\frac{1}{j\omega C_s}\right)\right) - I_s\left(\frac{j\omega M}{j\omega L_L + \left(\frac{1}{j\omega C_L} + R_L\right)}\right)(j\omega M) \tag{11}$$

The input impedance is calculated based on the simplified equivalent circuit model. The input impedance is given by [26]:

$$Z_s = \frac{V_s}{I_s} = \frac{((j\omega C_s)R_s + 1 - \omega^2 L_s C_s)((j\omega C_L)R_L + 1 - \omega^2 L_L C_L) - \omega^4 M^2 C_s C_L}{j\omega C_s((j\omega C_L)R_L + 1 - \omega^2 L_L C_L)} \tag{12}$$

The power transfer efficiency is given by ratio of output power to the input power:

$$\eta = \frac{P_{Out}}{P_{in}} = \frac{I_L^2 R_L}{I_s^2 Z_s} \tag{13}$$

For resonance coupling system, assuming $C = C_s = C_{L.}$

By substituting Eq. (10)–(12) in Eq. (13), the effective power efficiency is calculated as:

$$\eta = \left(\frac{j\omega M}{j\omega L_L + \left(\frac{1}{j\omega C_L} + R_L\right)}\right)^2 \frac{R_L}{\left(\left(R_s + j\omega L_s + \left(\frac{1}{j\omega C_s}\right) + \left(\frac{\omega^2 M^2}{j\omega L_L + \left(\frac{1}{j\omega C_L} + R_L\right)}\right)\right)\right)} \tag{14}$$

where, M is Mutual inductance between the coil antennas, L_s and L_L are the inductance of transmitter and receiver coils, C_s and C_L are matching capacitor for transmitter and receiver coils, R_s and R_L are internal resistance and load resistance of the coils and V_s and V_L are source and load voltages.

The transferred power is maximum at resonance frequency when load current in the circuit becomes maximum, at a given resonance frequency, Eq. (15) describes the condition for maximum power transfer efficiency [29]:

$$M^2 = \frac{R_L^2}{\omega_0^2} \tag{15}$$

The wireless power transfer efficiency using inductive coupling can be greater than 90% within a limited transmission range. During the last decade, methods such as magnetic resonance coupling for WPT have been widely studied by researchers to increase the efficiency of power transmission with greater distance range. Two-loop and four-loop coil systems were studied for magnetic resonance coupling based WPT system [30]. In order to achieve maximum power transfer efficiency, several studies have been done such as loop to coil coupling manipulation [29], automated impedance matching [31], adaptive frequency tuning [32], circuit structure manipulation [33], improving WPT for future portable consumer electronics using large transmitter coil system [34] and improving efficiency by four-coil system for deep brain simulation [35]. There are studies, which focused on misalignment between receiver and transmitter coil for applications such as wireless EV charging system [36] and wireless mobile phone charging system [37, 38]. With the rapid development in consumer electronics, there is a substantial increase in applications of NFC based wireless power transfer technology.

4. Eddy currents challenge: role of ferrite materials in NFC

As NFC technology relies on generated mutual inductance between the transmit and receive coil antennas, the amount of magnetic flux between them should be maximized to induce more current, thus increasing data transmission range. However, by placing a NFC tag on a metal surface, the efficiency of data transmission is greatly suppressed due to the generation of eddy currents within the metal surface [39]. To increase the efficiency and range while minimizing the losses due to eddy currents, a soft magnetic ferrite sheet can be inserted between the metal case and the antenna. **Figure 7(a)** illustrates how the magnetic field generates in NFC communication due to nearby conductive surface/plate, and **Figure 7(b)** shows the magnetic field produced when NFC area is shielded by a ferrite sheet from a conductive surface [40]. One can see the shielding effects due to insertion of ferrite sheet [40]. To be amenable to NFC, the ferrite sheets should have high permeability and low magnetic loss to concentrate the magnetic flux generated between the transmit and receive coils (antennas) [39, 41]. Ni-Zn ferrites and Mn-Zn ferrites are the most widely used soft magnetic materials for the preparation of these NFC ferrite sheets at high frequencies. Ni-Zn ferrites have exhibited higher operating frequency range up to 100 MHz as compared to Mn-Zn ferrites (a few MHz), which limits the use of Mn-Zn ferrites in NFC devices. Mn-Zn ferrites have been used in mini dc-dc converters, inductors and power inductors due to their high saturation induction and low losses [42]. For NFC applications, Ni-Zn ferrites offer better suited properties because of its high resistivity, high permeability, low magnetic loss, high operation frequency, and chemical stability. In particular, the high

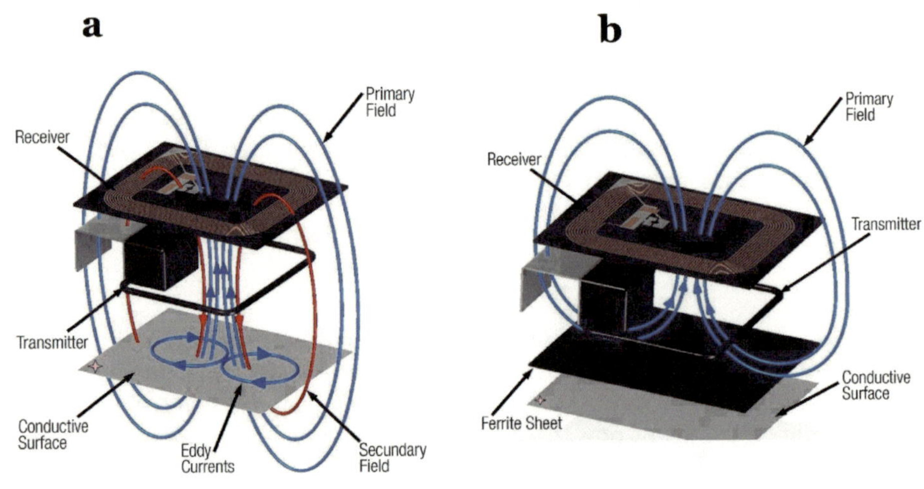

Figure 7.
(a) Eddy currents generated in NFC communication area due to conductive plate in vicinity; (b) magnetic field generated in NFC communication area with an incorporated ferrite sheet shielding [40].

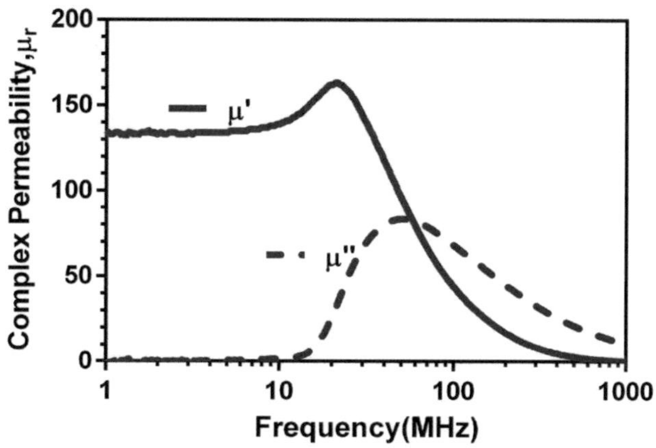

Figure 8.
A typical representation of relative permeability versus frequency of a ferrite sheet [45].

permeability and low magnetic loss of Ni-Zn ferrite sheets help to concentrate more magnetic flux and reduce eddy currents via magnetic shielding [42, 43].

To enhance the performance of NFC systems, the employment of high permeability and low loss magnetic sheets is highly desirable. The magnetic permeability and loss properties of Ni-Zn ferrites can be tailored by making strategic changes in crystallography, morphology and microstructure of the material. The relation between the complex relative permeability and frequency is termed as permeability dispersion. The frequency dependent relative permeability is given by Eq. (16) [44],

$$\mu_r = \mu' - j\mu'' \tag{16}$$

where μ_r is the ratio of the permeability of the material versus that of the free space (μ_0). μ' and μ'' are real and imaginary parts of the relative permeability, respectively.

The magnetic loss tangent is the ratio between the real and imaginary parts given by Eq. (17):

$$tan\ \delta_m = \frac{\mu''}{\mu'} \tag{17}$$

Figure 8 shows a typical relative permeability spectra of flexible ferrite sheet [45]. To achieve high signal transmission efficiency between NFC devices and increase the range of transmission, the relative permeability (μ') should be greater than 100 and the loss tangent ($tan\ \delta_m$) should be less than 0.05 at the standardized NFC operation frequency of 13.56 MHz [39]. Both μ' and μ'' of ferrite materials are greatly affected by the composition, microstructure and morphology, which are also quite sensitive to the processing parameters [46]. The key to achieve high performance ferrites for the targeted NFC applications is to tailor and optimize their synthesis process parameters. There are several different ways to synthesize Ni-Zn ferrites such as sol–gel method, citrate precursor method, hydrothermal synthesis and solid-state synthesis methods [42].

5. NFC modes of communication

NFC devices can communicate in either one of the two modes: active and passive mode. These modes determine how two NFC-enabled devices talk to each other. The distinction between modes depends on whether, a device generates its own RF field or used power from another device. In communication, initiator is the device that starts the communication, and target is the device that receives the signal from initiator. The main differences between main properties of passive technologies (NFC, Chipless RFID and UHF RFID) and active technologies (Bluetooth and Zigbee) are summarized in **Table 2** [47–49].

5.1 Active mode

In active mode, both NFC devices (initiator and target) send and receive data signals actively by using alternate RF (Radio frequency) field. Both NFC devices are self-powered and does not require to send power to target to perform the task, for example, devices such as smartphone or a self-powered tag. In active mode, the data

Feature	NFC	Bluetooth	UHF RFID	Chipless RFID	Zigbee
Read Range	1-2 cm for proximity cards with energy harvesting, 0.5 m for vicinity cards	10–100 m	Up to 15 m with inlay tags with 2 dBm read IC sensitivity, 3 m for UHF sensors with −9 dBm read sensitivity, 30 m BAP	<50 cm frequency coded and 2–3 m for time coded UWB	10–100 m
Memory capacity	<64 Kbytes	Several Kbytes depending on microcontroller	<64 Kbytes	<40 bytes	250–400 Kbytes
Energy source	Passive or semi-passive	Active	Passive or semi-passive	Passive	Active
Cost	Low	Low	Low	Moderate	Low
Universal frequency regulation	Yes	Yes	No	No	Yes
Security	High	Low	High	High	Low
Setup time	Less than 0.1 s	Approx. 6 s	Less than 0.1 s	—	Approx. 0.5 s
ID rewritable	Yes	Yes	Yes	No	Yes
Energy harvesting	Approx. 10 mW	No	Few μW	No	No
Reader cost	Low	Low	High	High	Low
Spectrum	13.56 MHz	2.4 GHz	433 MHz, 860–960 MHz	2.4–5.8 GHz	2.4 GHz (Globally), 915 MHz (as Z-Waves in U.S.) and 868 MHz (Europe)
Usability	Easy, human centric	Moderate, data centric	Easy, Data and human centric	Easy, Data and human centric	Easy, data centric

Table 2.
Comparison between different wireless technologies [47–49].

is sent between two devices using amplitude shift keying (ASK) i.e., the base RF field signal (13.56 MHz) is modulated with data using coding schemes (Miller and Manchester Coding). Data transfer rates are higher in this mode and it can work well at longer distances [50, 51].

5.2 Passive mode

In passive mode, the initiator sends the RF field to power the target. In turn, target used the RF field and sends back the stored data via a process called load modulation (Manchester coding) [52]. It is the most common mode for NFC, as it requires no battery and it is less expensive [53].

Three different combinations of communications are possible when two NFC device communicates with each other wirelessly, active-active, active-passive and passive-active. These are listed in **Table 3** [53].

While working in active and passive modes, the NFC devices perform different operation during communication. This means NFC device 1(initiator) must send signal first to NFC device 2 (target) to get the response back from device 2 (target). It is not possible for NFC device 2 (Target) to send data to device 1 without receiving any initial signal. All the possible interaction styles of NFC devices are listed in **Table 4** [14].

According to the NFC forum's device requirement, a device must have the functionality i.e., device needs to operate in reader/writer mode and in peer mode in order to be NFC-compliant [54]. i.e., a device must behave as an initiator during passive communication and an initiator or target during active communication. Initially, the NFC operating frequency of 13.56 MHz was unregulated. In 2004, NFC forum was established to standardize the tags and their operating protocols. There were three tasks standardized by the NFC Forum: including transferring power from a NFC device to a NFC tag, sending information from a NFC device to a NFC tag via signal modulation, and sensing the modulation by the load created on the NFC tag while performing load modulation to receive information from a NFC device. These three operation modes were designated by the NFC forum as reader/writer, peer-to-peer, and card emulation communications, as depicted in **Figure 9**. These are three main modes, under which a NFC device can operate [14]:

Device 1	Device 2	RF field generation
Active	Active	The RF field is generated by both devices
Active	Passive	The RF field is generated by device 1 only
Passive	Active	The RF field is generated by device 2 only

Table 3.
Various possible communication arrangements between two NFC devices [53].

Initiator device	Target device
NFC mobile	NFC tag
NFC mobile	NFC mobile
NFC reader	NFC mobile

Table 4.
Various possible interaction styles of NFC devices [14].

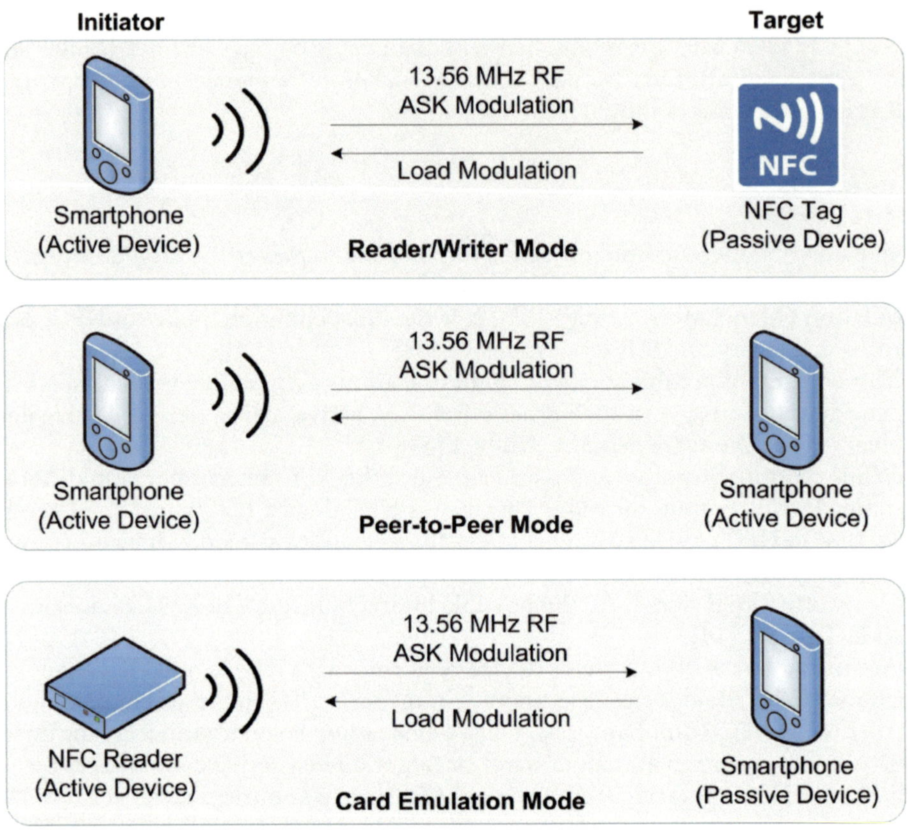

Figure 9.
Three different communication modes of a NFC device [14].

5.3 Reader/writer mode

In reader/writer mode, the device must be able to read and write with different types of NFC tags. This mode enables one NFC mobile to exchange data with one NFC tag. In reader/writer mode, most NFC devices act as readers and works in active mode to read the content of tag, such as contactless smart cards and RFID tags. In order to interact with tag appropriately, the device needs to detect the correct tag type. For that, the NFC device relies on an anticollision algorithm to select one tag when device comes across two tags simultaneously. An NFC device also works in writer mode, it can write data to the tags having writer application, such as TagWriter [55, 56]. In reader/writer mode, the NFC devices are compliant with ISO/IEC 14443A/B or Felica schemes tag types [14]. Some of the applications of reader/writer mode are smart posters, remote shopping, remote marketing, and so on [18].

5.4 Peer to peer mode

In peer-to-peer mode, two NFC specific devices can exchange data such as, pairing Bluetooth devices or WiFi link set-up, exchange business cards or text messages. This mode is standardized on the ISO/IEC 18092 NFCIP-1 standards. Both the devices operate in active mode during the communication and data is sent over a bi-directional half duplex channel, which means second device can only transmit data once first NFC device finish the transmission [14].

5.5 Card emulation mode

In card emulation mode, a NFC device behaves as an external reader traditional contactless smart card. This enables contactless payments through credit cards, debit cards, loyalty cards, etc. by using NFC device without changing the existing infrastructure. For example, NFC enabled mobile device can even store multiple contactless smart card applications in one phone. Card emulation mode supports ISO/IEC 14443 Type A and B, and Felica standards [14].

6. NFC tags

In an NFC system, there is always an element which functions as the receptor in passive mode, such as NFC tag. NFC tag, also known as the smart tag or information tag, is a small, printed circuit which act as a bit of storage memory along with a radio chip attached to an antenna [18]. It works in a passive mode, during which it does not have its own power source but uses power from the NFC device that communicates with it via magnetic induction. NFC tags have a few inches of working distance, NFC device must be very close to read the tag. NFC tags are used for a variety of applications in our day-to-day life, such as payments, launching websites, virtual visiting cards, lock/unlock doors, pet animals tagging, share photos, videos, and other information, etc. To ensure interoperability, a classification has been established for NFC tags by NFC-Forum that provides necessary specifications between different tag providers and the manufacturers of devices. Currently, there are five different types of NFC tags, depending on storage capacity, data transfer rate and read/write ability [1].

6.1 Type 1 NFC tags

Type 1 tags are based on standard ISO14443A with a memory of 96 bytes, expandable up to 2 Kbytes. The rate of data transfer is 106 Kbps and type 1 NFC tags have read/re-write capability.

6.2 Type 2 NFC tags

Like Type 1 tags, Type 2 tags are also based on ISO 14443A standard. It has a memory of 48 bytes, expandable up to 2 Kbytes. The rate of data transfer is 106 Kbps and type 2 NFC tags have read/re-write capability.

6.3 Type 3 NFC tags

Type 3 tags are Japanese Sony FeliCa standard (JIS X 6319–4). It has more memory and faster data transfer speed as compare to type 1 & 2 tags. The memory is 2 Kbyte, expandable up to 1 Mbyte with a transfer rate of 212 Kbps.

6.4 Type 4 NFC tags

Type 4 tags work on both ISO 14443 A & B communications. These are manufactured either in read only or read/re-write modes. Unlike other tags, a user cannot decide the mode. The memory is up to 32 Kbytes and, transmission rates are high; between 106 to 424 Kbps.

Type of NFC Tag	Standard	Memory	Data Transfer Rate	Data Capability	Anti-Collision	Available Market Products
Type 1	ISO 14443 A	96 bytes, expandable to 2 Kbytes	106 Kbps	Read-Write Read only	No	Innovision Topaz, Broadcom, BCM20203
Type 2	ISO 14443 A	48 bytes, expandable to 2 Kbytes	106 Kbps	Read-Write Read Only	Yes	NXP MIFARE Ultralight, NTAG203, NTAG 210, NTAG 212, NTAG 213/215/216, NTAG I^2C
Type 3	JIS X 6319–4	2 Kbytes, expandable to 1 Mbytes	212 Kbps, 424 Kbps	Read-Write Read Only	Yes	Sony FeliCa
Type 4	ISO 14443 A & B	Up to 32 Kbytes	106 to 424 Kbps	Read-Write Read Only (Factory Manufactured)	Yes	NXP DESFire, NXP SmartMX-JCOP,
Type 5	ISO 15693	Up to 64 Kbytes	26.48 Kbps	Read-Write Read Only	Yes	NXP ICODE Series

Table 5.
Summary of five different types of NFC tags [1, 54].

6.5 Type 5 NFC tags

NFC forum released type 5 tags recently in 2015 that is the newest NFC tag. It is based on ISO 15693. It has working range up to 1.5 m that allows the communication with RFID tags.

There are many factors, which decide the type of NFC tag used for a particular application, such as the type of the application, memory and transmission rate requirements, working distance, and cost involved, etc. Normally, an App needs to be installed on NFC devices, smart phone, or smart watch to use the NFC tags (e.g, Apple Pay and Google Pay for payments). In 2019, Ahold Delhaize, an European super market giant, enables its shelves with NFC-enabled electronics labels that allows shoppers to obtain detailed information of the product and add items to their cart for self-checkout using smart phones [57]. In 2020, Apple recently launched a new feature in its IOS operating system, "App Clip" that allows only clips/snippets of an App to do the communications with NFC tags without downloading the whole App [58]. **Table 5** summarizes the various features like standards, memory, data transmission rate, and so on, of five types of NFC tags along with their typical uses [1].

Apart from these five types of tags, there is a Type 6 NFC tags that are based on ISO 15693-3 standards and used to store NDEF messages or applications focused on identification cards [54]. It has memory capacity of 8 Kbytes and data transfer rate of 26.48 Kbps. The newly emerged 3-D printing technology has been exploited to develop new type of tags (e.g., Kovio's NFC Barcodes) [59]. New material and printing technologies can open endless opportunities in the area of NFC communications. In the last decade, there is dramatic increase in smart phones and tablets enabled with NFC function.

7. NFC applications

Many NFC applications have been developed since its inception and have become parts of our daily lives. Several of them are discussed below.

7.1 NFC mobile payments applications

With the technological advancement in the last decade, NFC enabled mobile devices are changing the way users receive data, make payments, and exchange information across devices all over the world. The advanced innovation has already been in use in Europe, Asia and North America because of powerful impact of influential mobile network operators (MNOs) in these parts of the world [60]. Different technologies such as RFID, contactless smart card, NFC, short message service (SMS), unstructured supplementary service data (USSD), wireless application protocol (WAP), interactive voice response (IVR), and so on, all contributed in the success of mobile payments. Presently, the integration of NFC technology in contactless mobile payments led to a tap-and-go tasks. In NFC enabled device, user needs to just touch or present phone to NFC enabled device and transfer or share data without any physical connection. For mobile payments, NFC has been set to be compatible with android, windows, and iOS operating system smartphones. Presently, there are many phones which are NFC compatible such as Samsung's Galaxy Series, Google's Nexus Series, and the iPhone. Some of the NFC payment applications for tap-and-go are Google Pay, Apple Pay, Android Pay, PayPal, Samsung Pay, Square Wallet, LifeLock Wallet and Visa payWave. NFC enabled mobile payment is reducing the need for physical form of payment between consumer and merchants. For example, mobile point of sale (mPOS) units are providing wireless devices to replace traditional cash registers and sale terminals [61]. These units are wire free and easy to install, for example, customers can buy apple products without going to cashier using mPOS device. Social shopping and, mobile wallets are some other examples of mobile payments. According to a recent report from Technavio, the global mobile payment market size is expected to grow at a CAGR (Compound Annual Growth Rate) of close to 36% by the end of 2021. A report from GATE Mobile Wallet Trends (Global Acceptance Transactions Engine) also highlighted that the number of mobile payment users were close to 2.1 billion in 2019 [62]. Some of the related work in mobile payment applications are utilization of electronic vouchers using offline NFC payment service [63], payment authorization process using secured system built on a service oriented architecture (SOA) [64], secure end-to-end NFC based mobile payments protocol for security purpose [65], software card emulation in NFC based contactless smart card system for security [66], and design and initial evaluation of a touch based remote grocery shopping process [67, 68]. **Figure 10** depicts the NFC payment transfer using mobile phone, the credit card transaction with virtual card stored in a distant location [69].

Figure 10.
NFC payment process using mobile phone; credit card transaction process with virtual card stored in a distant location [69].

7.2 Transportation and ticketing

Currently, around one third world population uses smartphone, and this number will go up significantly in years to come. Transport and ticketing are the well-known and promising applications of NFC system. Commuters around the world added adoption of mobility in all aspects of their transit such as journey planning, ticketing, information, and so on [70]. For purpose of transport, NFC forum has recognized three basic use cases, including connection, access, and transactions. The usage of NFC services in transport and ticketing depends on the consumers having NFC enabled mobile phones that are compatible with NFC standard ISO/IEC 14443 [71]. For example, a ticket for transit can easily download from NFC enabled kiosk using an NFC enabled smartphone and then smartphone can tap to a reader to gain the access to ticketing informations [72]. NFC tags can be easily embedded in posters, products, and maps etc., to provide transport related services to everyone. Smart posters containing these NFC tags provide a variety of information and/or links for transport service websites. Card emulation and card reader are the modes utilized in NFC enabled transit system [73]. **Figure 11** represents the evolution of consumer digital payments in public transport fare collection system using smartphone based NFC interface [74] . In past few years, NFC transport and ticketing applications have expanded in various aspects of the travel [54]. For examples, NFC forum and International Air Transport Association (IATA) jointly published the NFC reference guide for air travel [75]. Accenture's survey on public transport users reveals a big increase in use of smartphones for paperless travel and social media during the travel [76]. Virtual ticketing system and secure mCoupon protocol [77, 78], secure payment service by Smart Touch Project [79], an automated reservation and ticketing service for tourists, a system for car parking access, payment system for ticketing [80], offline Tapango system for electronic ticketing process including comparison with traditional paper ticketing process [81] are some of the prominent application of NFC that are used in travels and ticketing developed recently.

7.3 Healthcare applications

In last decade, various new materials and technologies including NFC based technology has gained ground in healthcare by adding convenience and providing efficient healthcare facilities [82–86]. According to a report issued by Transparency Market Research (TMR), healthcare is the fastest growing part of NFC, representing a CAGR of 20.4% [87, 88] . NFC provides user friendly benefits in healthcare e.g., secure physical access to buildings, medications and equipments, medical information, real time updates on patient care, medical alerts, home

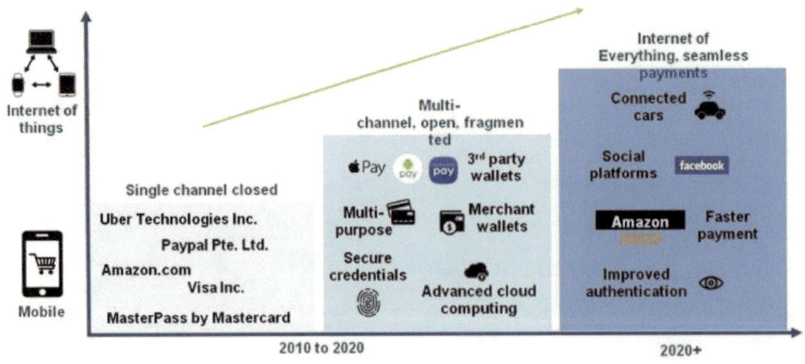

Figure 11.
The evolution of consumer digital payments including in public transport fare collection system using smartphone based NFC interface [74].

monitoring of patients, safer medications [89–92], storage of encrypted medical tags [93], adverse drugs and allergy detection system in hospitals [94] and electronic data recording services [95]. For example, recently Xiaomi launched Mi Band 5 wrist band, which supports NFC based payments and transactions through the band. Currently, in the time of COVID-19 pandemic, Silicon Craft Technology PLC (SICT) has launched an NFC enabled wearable band to track COVID-19 patients and those are under self-quarantine [96]. A major part of NFC in healthcare consists of health monitoring devices, such as NFC enabled blood pressure and activity monitors, wearable sensors [97, 98] and personal weight scale, etc., each of these devices send data to health centres connected via apps (TAPCheck blood pressure monitors, GENTAG, iMPAK Health for credit card-size RhythmTrak ECG device) [54]. There are more complex implementations of NFC technology in implantable health devices, such as heart monitors [99], cochlea implants [100–102], and optogenetics implants [103]. NFC enabled devices are also implanted in fitness and nutrition programs to promote overall health and wellness of users, such as, Apple watch, FitBit, Sony's Smart Band and Samsung Galaxy watch, and so on. There are affirmative studies that focus on improvement of health applications using NFC, such as self-diagnosis and medication, specific applications for the disabled, elderly and people with chronic diseases, etc. [49]. **Figure 12** shows few different applications of NFC in bio medical area [9, 98, 104, 105].

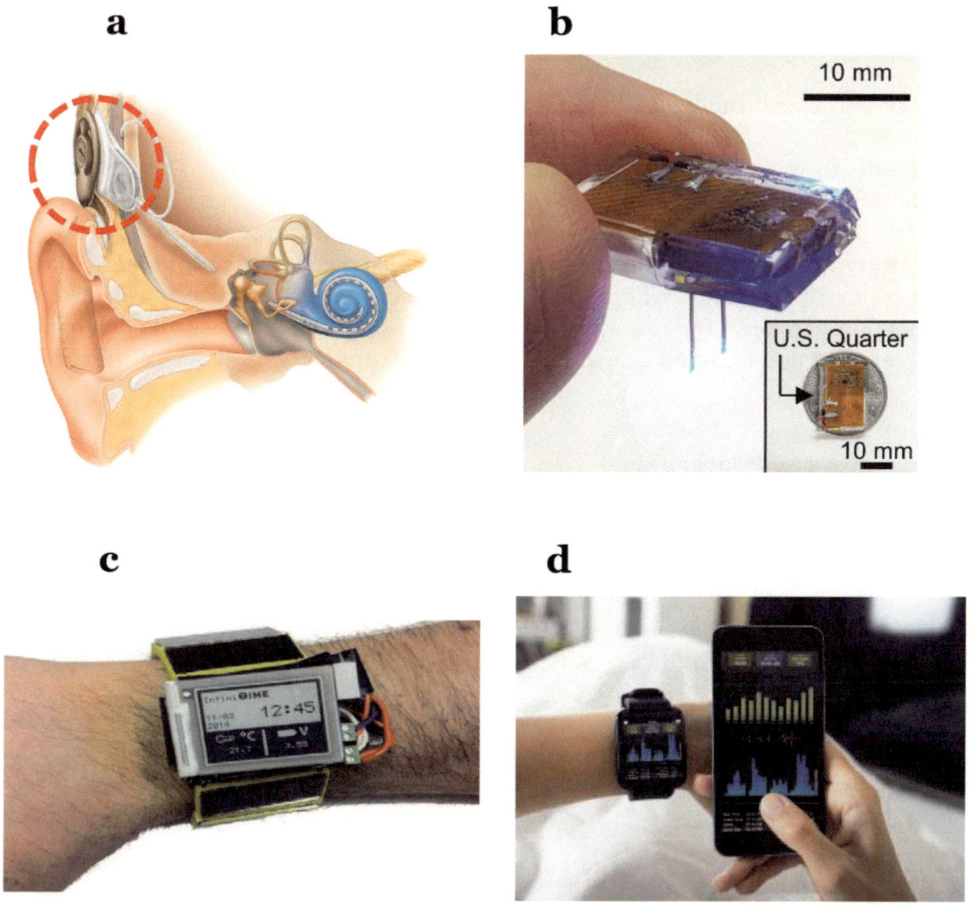

Figure 12.
Various healthcare applications of NFC; (a) cochlea implant with a circle showing the NFC communication part of the implant [9]; (b) Optogenetic implant for brain [104]; (c) wearable bracelet prototype [98]; (d) NFC enabled smart watch [105].

Reader/writer mode	Peer-to-peer mode	Card emulation mode
Increase mobility	Easy data exchange	Physical object elimination
Decrease physical effort	Device pairing	Access control
Ability to be adapted by many scenarios		
Easy to implement		

Table 6.
Benefits of NFC applications for each operating mode [14].

7.4 Access control and authentication

Access control services and authentication services using NFC technology has garnered much attention because of their initial promising results. The use of NFC enabled contactless smart cards for access, identification badges is similar to how users use the NFC enabled mobile phones to gain access in buildings. These types of applications indicates that NFC will play an important role in the next generation access control and identity management systems [54]. For example, two factor access control system for access in building using biometric fingerprint recognition for authentication and NFC for transferring the data to the computer controlled door [49]. Recently, the new 2021 BMW 5 Series launched NFC enabled digital car keys feature in their car which is compatible with Apple iPhone (Compatible with iPhone SE 2nd generation and iPhones with operating system above iOS 13) [96].

7.5 Gaming, social media and entertainment applications

The use of NFC enabled mobile games and social media has been brought to the market recently. In some of the NFC projects, gamers can tap each other phones together to receive access for new levels in the game and can score extra points. Similarly, social media users can establish social media connections and networking using peer-to-peer mode, where they can update their status and checked into the location also. Some of the examples are, Pass the bomb and Exquisite Touch games [106], smart phones are NFC enabled musical instruments [107], and Whack-a-Mole game [108].

7.6 Inventory and packaging

RFID and NFC enabled devices has been used in warehouses and stores for inventory control, labelling and packaging purpose. For example, NFC-enabled temperature sensors in tracking the food and drinks conditions, NFC tags for luxury products, NFC tags for authentication of instruments and linked to database to keep record of their repair date, NFC labels for food tracking (expiration date, gluten free, low calorie etc) [54].

Some of the other applications of NFC are location-based tracking, educational, work force and retail management applications. **Table 6** summarizes the benefits of each mode of NFC communications.

8. Privacy and security

Along with benefits, any new technology comes with new challenges and concerns. Similarly, NFC also has its own concerns, specifically the privacy and security concerns.

The data transfer in NFC technology relies on the magnetic flux strength, which is inversely proportional to the square of the distance between the NFC device and the tag antenna. Hence, magnetic flux strength decreases quickly as the distance between the NFC device and the tag antenna increases, thus restricting the range and effectiveness of NFC technology up to a few centimeters only. This disadvantage turns out be a great advantage and a flagship selling point for NFC technology as far as security is concerned. Along with the numerous benefits of NFC technology, researchers and technologists also are aware of the security and privacy concerns that comes along with technology. As this technology is managing our payment system and private information, the technology needs to be secure and safe from any kind of security threat [94, 109, 110]. Although NFC technology is based on contactless smartcards and payment technology and communicate in short range distance, yet there are chances of eavesdropping on communications [111]. Several types of attacks have been discussed in studies such as eavesdropping, data corruption or modification, relay attacks [112], man-in-the-middle attacks and DOS (Denial-of-Service). With an increase of NFC products in the market, the individual makers of product do not implement NFC technology perfectly. According to head of HP's Zero Day, in an annual hacking competition (Pwn2Own) in 2015, researchers used flaws of technology to compromise the NFC devices [113]. Several researchers tried to increase the range of NFC communication via different design topologies. For example, Range extension attacks on contactless smartcards [114], increase in range by placing metal plate under antenna, position and alignment of antenna to extend the range with a range extension to 13.4 cm by changing the antenna [115]. Many studies indicate that the usage of secure channels and examining the RF field can address most of these threats [116], while the high speed of data transfer and close distance handles the rest [117]. The close operating distance for NFC prevents an intruding signal from interrupting the signal or inserting threating data during the transfer [118, 119]. Some of the steps can help you safely use NFC technology for your mobile phone payments and other applications [113], including:

- Customers should read the fine print before using any NFC-enabled applications.
- Patch your device rapidly.
- Active your device only when it is in use.

9. Conclusion and future directions

In this chapter, we have covered an overview of NFC technology, including its current uses. NFC is an emerging technology and finding applications in every aspects of the daily life. In last decade, there has been explosion on both fronts, market applications and research. During the future years, it will expand further. NFC has potential to make personalized medicine and next generation point diagnosis at cellular and molecular levels realities. There is ongoing research to make NFC technology more affordable, easy to use, as well as more compact in size. There is a strong desire to improve the efficiency and wireless power transfer rates of these NFC devices. New materials research combined with evolving 3-D printing could led to new type of NFC tags, devices, and applications.

Presently, most NFC applications involve the use of mobile phones and exchange of sensitive personal financial and other data during the transitions. Most users have vast amount of other important data as well. This raises the concerns about privacy and security like spamming, unwanted contents on phones and so on. User privacy should be given top priority as NFC technology expands into new avenues. Cautions should be used while implementing NFC services. Data encryption, establishing secure channels between NFC devices and users' proper educations (locking code for the phone, regular update of antivirus software, erasing the phone in case its stolen) will be some key features in addressing the users' concerns about privacy and security. Overall, NFC is an exciting new technology that will present a wide variety of new applications along with renewed challenges in the years to come.

Author details

Poonam Lathiya[1,2*] and Jing Wang[1,2*]

1 Department of Electrical Engineering, University of South Florida, Tampa, FL, United States

2 Wireless and Microwave Information Systems and RF MEMS Transducers Lab, Tampa, FL, United States

*Address all correspondence to: poonam2@usf.edu and jingw@usf.edu

DOI: http://dx.doi.org/10.5772/intechopen.96345

References

[1] Coskun, V., B. Ozdenizci, and K.J.W. p.c. Ok, *A survey on near field communication (NFC) technology*. 2013. **71**(3): p. 2259–2294.

[2] Blog, T. *The Evolution of Near Field Communication*. Available from: https://www.techpats.com/evolution-near-field-communication-nfc/

[3] Bae, K.-J., et al., *The ubiquitous library for the blind and physically handicapped —A case study of the LG Sangnam Library, Korea*. 2007. **33**(3): p. 210–219.

[4] Exchange, T. *innovative examples of NFC technology*. January 22, 2013; Available from: https://exchange.telstra.com.au/6-innovative-examples-of-nfc-technology/.

[5] Hamblen, M. *A short history of NFC*. 2012; Available from: https://www.computerworld.com/article/2493888/a-short-history-of-nfc.html.

[6] Bite, B. *NFC Usage and Statistics for 2020*. November 13, 2020; Available from: https://www.bluebite.com/nfc/the-state-of-nfc-in-2020

[7] Report, M.A., *The exponential growth of mobile internet application and advancement of 3G and 4G networks is anticipated to drive the market*. October, 2016.

[8] Want, R.J.I.p.c., *An introduction to RFID technology*. 2006. **5**(1): p. 25–33.

[9] Kim, H.-J., et al., *Review of near-field wireless power and communication for biomedical applications*. 2017. **5**: p. 21264–21285.

[10] Landt, J.J.I.p., *The history of RFID*. 2005. **24**(4): p. 8–11.

[11] Atzori, L., A. Iera, and G. Morabito, *The Internet of Things: A survey Comput. Netw*. 2010.

[12] ZARE, M.Y., *RFID: a bibliographical literature review with future research directions*. 2014.

[13] Arcese, G., et al., *Near field communication: Technology and market trends*. 2014. **2**(3): p. 143–163.

[14] Coskun, V., K. Ok, and B. Ozdenizci, *Near field communication (NFC): From theory to practice*. 2011: John Wiley & Sons.

[15] Suparta, W.J.J.o.C.S., *Application of near field communication technology for mobile airline ticketing*. 2012. **8**(8): p. 1235.

[16] Sabella, R.R. *Near Field Communication*. Available from: https://www.dummies.com/consumer-electronics/needed-near-field-communication/.

[17] Moon, W.Y. and S.D. Kim. *A payment mediation platform for heterogeneous FinTech schemes*. in *2016 IEEE Advanced Information Management, Communicates, Electronic and Automation Control Conference (IMCEC)*. 2016. IEEE.

[18] Curran, K., et al., *Near Field Communication*. 2012. **2**(3).

[19] Huang, H.-P., C.-S. Chen, and T.-Y. Chen. *Mobile diagnosis based on RFID for food safety*. in *2006 IEEE International Conference on Automation Science and Engineering*. 2006. IEEE.

[20] Reinhold, C., et al., *Efficient antenna design of inductive coupled RFID-systems with high power demand*. 2007. **2**(6): p. 14–23.

[21] Finkenzeller, K., *RFID handbook: fundamentals and applications in contactless smart cards, radio frequency identification and near-field communication*. 2010: John wiley & sons.

[22] Atluri, S. and M. Ghovanloo. *Design of a wideband power-efficient inductive wireless link for implantable biomedical devices using multiple carriers*. in *Conference Proceedings. 2nd International IEEE EMBS Conference on Neural Engineering, 2005*. 2005. IEEE.

[23] Hannakam, L.J.A.f.E., *Berechnung der gegeninduktivität achsenparalleler zylinderspulen*. 1967. **51**(3): p. 141–154.

[24] Yee, K.J.I.T.o.a. and propagation, *Numerical solution of initial boundary value problems involving Maxwell's equations in isotropic media*. 1966. **14**(3): p. 302–307.

[25] Warnick, K.F., et al., *Optimizing power transfer efficiency and bandwidth for near field communication systems*. 2012. **61**(2): p. 927–933.

[26] Kong, S., et al. *Analytical expressions for maximum transferred power in wireless power transfer systems*. in *2011 IEEE International Symposium on Electromagnetic Compatibility*. 2011. IEEE.

[27] Shen, F., et al. *Circuit analysis of wireless power transfer by "coupled magnetic resonance"*. in *IET International Communication Conference on Wireless Mobile and Computing (CCWMC 2009)*. 2009. IET.

[28] Liu, X., et al., *Analysis of efficiency improvement in wireless power transfer system*. 2017. **11**(2): p. 302–309.

[29] Agcal, A., S. Ozcira, and N. Bekiroglu, *Wireless power transfer by using magnetically coupled resonators*, in *Journal of Wireless Power Transfer: Fundamentals and Technologies*. 2016. p. 49–66.

[30] Dang, Z., *Magnetic resonance coupled wireless power transfer systems*. 2013, University of Alabama Libraries.

[31] Beh, T.C., et al., *Automated impedance matching system for robust wireless power transfer via magnetic resonance coupling*. 2012. **60**(9): p. 3689–3698.

[32] Sample, A.P., D.T. Meyer, and J.R.J. I.T.o.i.e. Smith, *Analysis, experimental results, and range adaptation of magnetically coupled resonators for wireless power transfer*. 2010. **58**(2): p. 544–554.

[33] Chen, L., et al., *An optimizable circuit structure for high-efficiency wireless power transfer*. 2011. **60**(1): p. 339–349.

[34] Hoang, H., et al., *An adaptive technique to improve wireless power transfer for consumer electronics*. 2012. **58** (2): p. 327–332.

[35] Yang, C.-L., et al., *Efficient four-coil wireless power transfer for deep brain stimulation*. 2017. **65**(7): p. 2496–2507.

[36] Moon, S. and G.-W.J.I.T.o.P.E. Moon, *Wireless power transfer system with an asymmetric four-coil resonator for electric vehicle battery chargers*. 2015. **31** (10): p. 6844–6854.

[37] Zhong, W., C.K. Lee, and S.R.J.I.t.o. i.e. Hui, *General analysis on the use of Tesla's resonators in domino forms for wireless power transfer*. 2011. **60**(1): p. 261–270.

[38] Huang, M., Y. Lu, and R.P.J.I.T.o.P. E. Martins, *A reconfigurable bidirectional wireless power transceiver for battery-to-battery wireless charging*. 2018. **34**(8): p. 7745–7753.

[39] Lee, W., et al., *A simple wireless power charging antenna system: Evaluation of ferrite sheet*. IEEE Transactions on Magnetics, 2017. **53**(7): p. 1–5.

[40] Victoria, J., et al., *Improving the Efficiency of NFC Systems Through Optimizing the Sintered Ferrite Sheet Thickness Selection*. 2020. **62**(4): p. 1504–1514.

[41] Lee, B., et al., *NFC antenna design for low-permeability ferromagnetic material.* EEE Antennas Wireless Propagation Letters, 2014. **13**: p. 59–62.

[42] Tsutaoka, T., *Frequency dispersion of complex permeability in Mn–Zn and Ni–Zn spinel ferrites and their composite materials.* Journal of Applied Physics, 2003. **93**(5): p. 2789–2796.

[43] Wu, X., et al., *Influence of particle size on the magnetic spectrum of NiCuZn ferrites for electromagnetic shielding applications.* Journal of Magnetism Magnetic Materials, 2016. **401**: p. 1093–1096.

[44] Kumar, S., T. Shinde, and P. Vasambekar, *Engineering High Permeability: M n–Z n and N i–Z n Ferrites.* International Journal of Applied Ceramic Technology, 2015. **12**(4): p. 851–859.

[45] Lathiya, P., M. Kreuzer, and J. Wang, *RF complex permeability spectra of Ni-Cu-Zn ferrites prepared under different applied hydraulic pressures and durations for wireless power transfer (WPT) applications.* Journal of Magnetism and Magnetic Materials, 2020. **499**: p. 166273.

[46] Lathiya, P. and J. Wang, *Effects of the Sintering Temperature on RF Complex Permeability of NiCuCoZn Ferrites for Near-Field Communication Applications.* IEEE Transactions on Magnetics, 2019. **55**(2): p. 1–4.

[47] Lazaro, A., et al., *NFC Sensors Based on Energy Harvesting for IoT Applications*, in *Recent Wireless Power Transfer Technologies*. 2019, IntechOpen.

[48] Cao, Z., et al., *Near-field communication sensors.* 2019. **19**(18): p. 3947.

[49] Coskun, V., B. Ozdenizci, and K.J.S. Ok, *The survey on near field communication.* 2015. **15**(6): p. 13348–13405.

[50] Mareli, M., et al. *Experimental evaluation of NFC reliability between an RFID tag and a smartphone*. in *2013 Africon*. 2013. IEEE.

[51] Paus, A.J.C.f.C.S., *Near field communication in cell phones.* 2007. **24**(8).

[52] SERIALIO.COM. *The NFC Forum Standard*. 1996–2020; Available from: https://www.serialio.com/support/learn-rfid/what-near-field-communication-nfc#:~:text=The%20NFC%20device%20in%20Reader,magnetic%20field%20from%20the%20reader.

[53] Haselsteiner, E. and K. Breitfuß. *Security in near field communication (NFC)*. in *Workshop on RFID security*. 2006. sn.

[54] McHugh, S. and K.J.S.L.o.E.T.i.L. Yarmey, *Near field communication: recent developments and library implications.* 2014. **1**(1): p. 1–93.

[55] Sabella, R.R. *The NFC Operating Modes*. Available from: https://www.dummies.com/consumer-electronics/the-nfc-operating-modes/.

[56] Forum, N. *he Operating Modes Of NFC Devices*. December 17, 2013]; Available from: https://nfc-forum.org/resources/what-are-the-operating-modes-of-nfc-devices/.

[57] Clark, M. *Ahold Delhaize to roll out NFC shelf edge labels in supermarkets across Europe*. September 16, 2019; Available from: https://www.nfcw.com/2019/09/16/364378/ahold-delhaize-to-roll-out-nfc-shelf-edge-labels-in-supermarkets-across-europe/

[58] O'Boyle, B. *What are App Clips on iPhone and how do they work?* December 17, 2020; Available from: https://www.pocket-lint.com/phones/news/apple/152664-what-are-app-clips-on-iphone-and-how-do-they-work.

[59] Verimatrix. *Inside Secure NFC SOLUTIONS NOW KOVIO RF*

BARCODE READY REVOLUTIONARY "PRINTED SILICON" TAGS ENABLE ITEM-LEVEL INTERACTION BETWEEN CONSUMERS AND BRANDS. 2012; Available from: https://www.insidesecure.com/Company/Press-releases/Inside-Secure-NFC-SOLUTIONS-NOW-KOVIO-RF-BARCODE-READY-REVOLUTIONARY-PRINTED-SILICON-TAGS-ENABLE-ITEM-LEVEL-INTERACTION-BETWEEN-CONSUMERS-AND-BRANDS.

[60] Ondrus, J. and Y. Pigneur. *An assessment of NFC for future mobile payment systems*. in *International Conference on the Management of Mobile Business (ICMB 2007)*. 2007. IEEE.

[61] Alliedwallet. *The Future of Mobile Payment Technology*. 2020; Available from: https://www.alliedwallet.com/blog/blog-posts/future-mobile-payment-technology/.

[62] Blog, T. *The Future Trends of Mobile Payment: NFC Payments to Expand its Majority Market Share*. August 9, 2019; Available from: https://blog.technavio.com/blog/mobile-payment-trends-nfc-payments-leads-growth (13).

[63] Van Damme, G., et al. *Offline NFC payments with electronic vouchers*. in *Proceedings of the 1st ACM workshop on Networking, systems, and applications for mobile handhelds*. 2009.

[64] Kadambi, K.S., J. Li, and A.H. Karp. *Near-field communication-based secure mobile payment service*. in *Proceedings of the 11th international Conference on Electronic Commerce*. 2009.

[65] Bojjagani, S., V.J.C.S. Sastry, and Interfaces, *A secure end-to-end proximity NFC-based mobile payment protocol.* 2019. **66**: p. 103348.

[66] Roland, M. *Software card emulation in NFC-enabled mobile phones: great advantage or security nightmare*. in *Fourth international workshop on security and privacy in spontaneous interaction and mobile phone use*. 2012.

[67] Cappiello, I., S. Puglia, and A. Vitaletti. *Design and initial evaluation of a ubiquitous touch-based remote grocery shopping process*. in *2009 First International Workshop on Near Field Communication*. 2009. IEEE.

[68] TouchDynamic. *Mobile Payment With Nfc Near Field Communication Technology*. 2021; Available from: https://www.touchdynamic.com/contactless-mobile-payments-are-they-really-a-good-idea/mobile-payment-with-nfc-near-field-communication-technology/.

[69] Manjunatha, P. *How does Mobile Payment App Work?* 2021; Available from: https://www.techsagar.com/how-does-mobile-payment-app-work/.

[70] Rao, N. *The Increasing Adoption of NFC in Public Transportation*. March 2, 2018; Available from: https://blog.sasken.com/the-increasing-adoption-of-nfc-in-public-transportation.

[71] Brumercikova, E., B. Bukova, and L. J.C.-S.l.o.t.U.o.Z. Krzywonos, *NFC technology in public transport.* 2016. **18** (2): p. 20–25.

[72] Burden, M. *Near field communications (NFC) in public transport*. in *The Instititon of Engineering and Technology Seminar on RFID and Electronic Vehicle Identification in Road Transport*. 2006. IET.

[73] Chandrasekar, P. and A.J.W.P.C. Dutta, *Recent Developments in Near Field Communication: A Study.* 2020: p. 1–20.

[74] Report, G.M.T. *Transit Ticketing & Fare Collection in APAC*. October 1, 2017; Available from: https://www.globalmasstransit.net/archive.php?id=27988.

[75] IATA, N.F.a. *NFC Reference Guide for Air Travel.* . 2013; Available from: https://www.iata.org/whatwedo/stb/fast-travel/Documents/nfc-reference-guide-air-travel.pdf. 28.

[76] Accenture. May 21, 2013; Available from: http://newsroom.accenture.com/news/public-transportation-users-predict-big-increases-in-the-use-ofsmartphones-paperless-travel-and-social-media-new-accenture-survey-reveals.htm. 28

[77] Hsiang, H.-C., et al., *A secure mCoupon scheme using near field communication.* 2009. **5**(11): p. 3901–3909.

[78] Aigner, M., S. Dominikus, and M. Feldhofer. *A system of secure virtual coupons using NFC technology*. in *Fifth Annual IEEE International Conference on Pervasive Computing and Communications Workshops (PerComW'07)*. 2007. IEEE.

[79] Pasquet, M., J. Reynaud, and C. Rosenberger. *Secure payment with NFC mobile phone in the SmartTouch project*. in *2008 International Symposium on Collaborative Technologies and Systems*. 2008. IEEE.

[80] Baldo, D., G. Benelli, and A. Pozzebon. *The SIESTA project: Near Field Communication based applications for tourism*. in *2010 7th International Symposium on Communication Systems, Networks & Digital Signal Processing (CSNDSP 2010)*. 2010. IEEE.

[81] Neefs, J., et al. *Paper ticketing vs. Electronic Ticketing based on off-line system'Tapango'*. in *2010 Second International Workshop on Near Field Communication*. 2010. IEEE.

[82] Bazard, P., et al., *Plasmonic stimulation of electrically excitable biological cells*. 2018, Google Patents.

[83] Chen, W., et al., *Nanomachines and other caps on mesoporous silica nanoparticles for drug delivery.* Accounts of Chemical Research, 2019. **52**(6): p. 1531–1542.

[84] Chik, M.W., et al., *Polymer-wrapped single-walled carbon nanotubes: A transformation toward better applications in healthcare.* Drug Delivery Translational Research, 2019. **9**(2): p. 578–594.

[85] Damnjanovic, R., et al., *Hybrid Electro-Plasmonic Neural Stimulation with Visible-Light-Sensitive Gold Nanoparticles.* ACS Nano, 2020. **14**(9): p. 10917–10928.

[86] Lin, R., et al., *Wireless battery-free body sensor networks using near-field-enabled clothing.* Nature Communications, 2020. **11**(1): p. 1–10.

[87] Ardila, C. *Six ways NFC Helps Healthcare*. November 18, 2015; Available from: https://www.nxp.com/company/blog/six-ways-nfc-helps-healthcare:BL-6-WAYS-NFC-HELPS-HEALTHCARE.

[88] Research, T.M. *INCREASED DEMAND FOR SAFETY AND SECURITY BOOSTS ACCESS CONTROL AND AUTHENTICATION MARKET*. August 01, 2017.

[89] Fontecha, J., et al. *An NFC approach for nursing care training*. in *2011 Third International Workshop on Near Field Communication*. 2011. IEEE.

[90] Marcus, A., et al. *Using NFC-enabled mobile phones for public health in developing countries*. in *2009 First International Workshop on Near Field Communication*. 2009. IEEE.

[91] Iglesias, R., et al. *Experiencing NFC-based touch for home healthcare*. in *Proceedings of the 2nd international conference on pervasive technologies related to assistive environments*. 2009.

[92] Bravo, J., et al. *Touch-based interaction: An approach through NFC*. in

2007 3rd IET International Conference on Intelligent Environments. 2007. IET.

[93] Dunnebeil, S., et al. *Encrypted NFC emergency tags based on the German telematics infrastructure*. in *2011 Third International Workshop on Near Field Communication*. 2011. IEEE.

[94] Jara, A.J., et al. *Drugs interaction checker based on IoT*. in *2010 Internet of Things (IOT)*. 2010. IEEE.

[95] Morak, J., et al. *Near field communication technology as the key for data acquisition in clinical research*. in *2009 First International Workshop on Near Field Communication*. 2009. IEEE.

[96] Hegde, A. *NFC Latest Trends 2021: 7 NFC Technology Trends to Watch Out for!* December 28, 2020; Available from: https://blog.beaconstac.com/2020/10/nfc-latest-trends/.

[97] Zhan, B., et al., *Wearable near-field communication antennas with magnetic composite films.* 2017. 7(6): p. 065313.

[98] Magno, Michele, et al. "Infinitime: A multi-sensor energy neutral wearable bracelet." International Green Computing Conference. IEEE, 2014.

[99] Jara, A.J., et al. *Heart monitoring system based on NFC for continuous analysis and pre-processing of wireless vital signs*. in *Proc. Int. Conf. Health Informatics (HEALTHINF)*. 2012.

[100] Yip, M., A. Chandrakasan, and K. Stankovic, *Low power cochlear implants*. 2018, Google Patents.

[101] Erráez Castelltort, G., *Design of an electrical nerve stimulator using wireless power transmission through NFC*. 2018, Universitat Politècnica de Catalunya.

[102] Bazard, P., et al., *Nanoparticle-based plasmonic transduction for modulation of electrically excitable cells.* 2017. 7(1): p. 1–13.

[103] Biswas, D.K., et al. *An NFC (near-field communication) based wireless power transfer system design with miniaturized receiver coil for optogenetic implants*. in *2018 Texas Symposium on Wireless and Microwave Circuits and Systems (WMCS)*. 2018. IEEE.

[104] Kim, Choong Yeon, et al. "Soft subdermal implant capable of wireless battery charging and programmable controls for applications in optogenetics." Nature communications 12.1 (2021): 1-13.

[105] Gutirrez, L. *Using Smartphones as a Medical Device for Point-of-Care Applications*. 2020; Available from: https://starfishmedical.com/blog/smartphones-as-a-medical-device/.

[106] Nandwani, A., P. Coulton, and R. Edwards. *NFC mobile parlor games enabling direct player to player interaction*. in *2011 Third International Workshop on Near Field Communication*. 2011. IEEE.

[107] Ivanov, R. *Indoor navigation system for visually impaired*. in *Proceedings of the 11th International Conference on Computer Systems and Technologies and Workshop for PhD Students in Computing on International Conference on Computer Systems and Technologies*. 2010.

[108] Broll, G., et al. *Touch to play: mobile gaming with dynamic, NFC-based physical user interfaces*. in *Proceedings of the 12th international conference on Human computer interaction with mobile devices and services*. 2010.

[109] Schoo, P. and M. Paolucci. *Do you talk to each poster? Security and Privacy for Interactions with Web Service by means of Contact Free Tag Readings*. in *2009 First International Workshop on Near Field Communication*. 2009. IEEE.

[110] Jara, A.J., et al. *Evaluation of the security capabilities on NFC-powered devices*. in *European Workshop on Smart*

Objects: Systems, Technologies and Applications. 2010. VDE.

[111] Kennedy, T. and R. Hunt. *A review of WPAN security: attacks and prevention*. in *Proceedings of the international conference on mobile technology, applications, and systems*. 2008.

[112] Francis, L., et al. *Practical NFC peer-to-peer relay attack using mobile phones*. in *International Workshop on Radio Frequency Identification: Security and Privacy Issues*. 2010. Springer.

[113] Lemos, R. *NFC security: 3 ways to avoid being hacked*. June 26, 2015; Available from: https://www.pcworld.com/article/2938520/nfc-security-3-ways-to-avoid-being-hacked.html

[114] Oren, Y., D. Schirman, and A. Wool. *RFID jamming and attacks on Israeli e-voting*. in *Smart SysTech 2012; European Conference on Smart Objects, Systems and Technologies*. 2012. VDE.

[115] Hermans, B. and S. Moeniralam, *Extending the range of NFC capable devices*. 2017.

[116] Medaglia, C.M. and A. Serbanati, *An overview of privacy and security issues in the internet of things*, in *The internet of things*. 2010, Springer. p. 389–395.

[117] Ling, J., Y. Wang, and W.J.I.N.S. Chen, *An Improved Privacy Protection Security Protocol Based on NFC*. 2017. **19** (1): p. 39–46.

[118] Jayapandian, N., *Business Transaction Privacy and Security Issues in Near Field Communication*, in *Network Security and Its Impact on Business Strategy*. 2019, IGI Global. p. 72–90.

[119] Eun, H., H. Lee, and H.J.I.T.o.C.E. Oh, *Conditional privacy preserving security protocol for NFC applications*. 2013. **59**(1): p. 153–160.

Chapter 7

WPT, Recent Techniques for Improving System Efficiency

Mohamed Aboualalaa, Hala Elsadek and Ramesh K. Pokharel

Abstract

Wireless power transfer (WPT) technologies have received much more attention during the last decade due to their effectiveness in wireless charging for a wide range of electronic devices. To transmit power between two points without a physical link, conventional WPT systems use two coils, one coil is a transmitter (Tx) and the other is a receiver (Rx) which generates an induced current from the received power. Two main factors control the performance of the WPT schemes, power transfer efficiency (PTE) and transmission range. Power transfer efficiency refers to how much power received by the rechargeable device compared to the power transmitted from the transmitter; while transmission range indicates the longest distance between transmitter and receiver at which the receiver can receive power within the acceptable range of power transfer efficiency. Several studies were carried out to improve these two parameters. Many techniques are used for WPT such as inductive coupling, magnetic resonance coupling, and strongly coupled systems. Recently, metamaterial structures are also proposed for further transfer efficiency enhancement. Metamaterials work as an electromagnetic lensing structure that focuses the evanescent transmitted power into receiver direction. Transmitting & Receiving antenna systems may be used for sending power in certain radiation direction. Optimizing the transmitter antenna and receiver antenna characteristics increase the efficiency for WPT systems. This chapter will present a survey on different wireless power transmission schemes.

Keywords: capacitive coupling, inductive coupling, intermediate resonators, magnetic resonance coupling, metamaterial structures, power transfer efficiency (PTE), strongly coupled magnetic resonance, wireless power transmission (WPT)

1. Introduction

With the spreading of mobile phones, portable and wearable electronic devices and changes in the human lifestyle, the need for WPT technology grows to get rid of the inconvenience due to using power cables. On the other hand, there are some applications where WPT probably the only solution or the most efficient solution for their powering for instance implanted biomedical devices, buried sensors, some sensors found in a severe environment such as very high temperatures, and so forth. One of the first trials for WPT was performed by Nikola Tesla a century ago. He wanted to develop a wireless power distribution system. **Figure 1** illustrates a simplified diagram of a WPT system which simply consists of a transmitter that sends the transmitted power through an RF coil or RF resonator. On the receiver

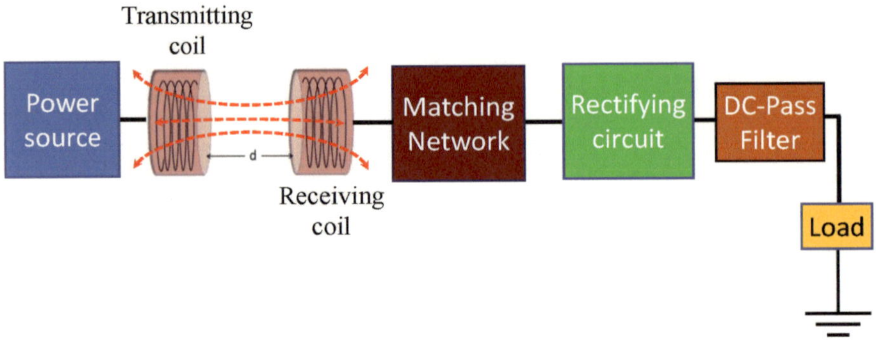

Figure 1.
WPT system.

side, there is a receiving resonator which can be an antenna or coil to receive the incoming wave from the transmitter. Afterward, an impedance matching circuit is inserted to ensure maximum power transfer between the receiving resonator and the rectifying circuit. Then, the rectifying stage is connected. Many combinations could be used for the rectification purpose such as half-wave, full-wave, or any series/parallel diodes combinations. All these rectification circuits are used for converting RF power into DC power. In order to achieve smoothing DC output voltage as well as blocking the higher-order modes, the rectifying circuit is followed by a DC pass filter. The final stage is the device (load) that needs to be charged wirelessly. In this chapter, we will focus on the coupled resonators which is the first stage for WPT systems.

Wireless power transfer technologies can be divided into different categories such as inductive coupling, resonant inductive coupling, capacitive coupling,

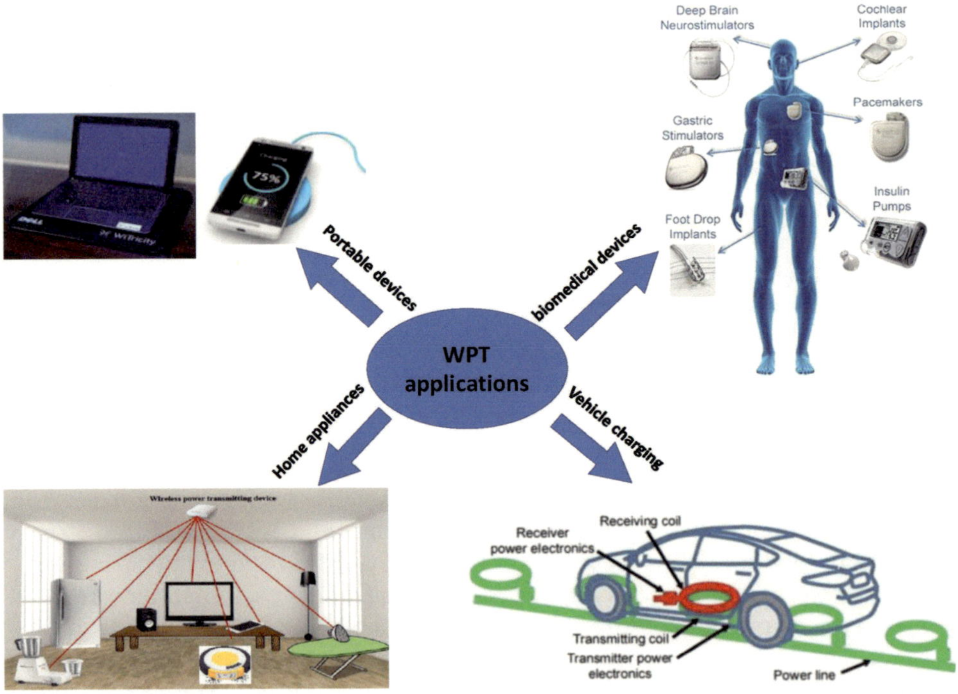

Figure 2.
WPT applications.

microwaves. Through this chapter, we will cover these technologies with highlights on the recent techniques for improving the power transfer efficiency such as using intermediate resonators, applying metasurface structures, and so on. **Figure 2** shows the current and potential applications for WPT systems.

2. Inductive coupling WPT

Conventional coils of wire are the simplest way to transmit a wireless power between transmitter and receiver. In this case, the system can be represented as a transformer where a transmitting coil is analogous to the primary coil, while the received coil is equivalent to the secondary coil as revealed in **Figure 3**. An inductive power transfers between the two coils in a form of a magnetic field. The intensity of the magnetic field follows Ampere's law as in (1), where $\bar{H}$ is the magnetic field intensity that is generated when an electric current, I, passes through an electric closed path with a length of l.

$$\oint \bar{H}.dl = I \tag{1}$$

When the Transmitter has a time-varying current and mounted at an appropriate position from the receiver. Receiver's coil cuts the magnetic field lines, and an induced electromotive force (emf) is generated between the terminals of the receiver's coil as shown in **Figure 3**. The value of the emf depends on the time-varying of the magnetic flux (ϕ) as characterized by Faraday's law as in (2). It is clear that this WPT technology is valid only for short-range applications for example wireless charging pads to recharge cellphones and handheld wireless devices such as laptops and tablets, electric toothbrush, shaver's battery charging, induction stovetops and industrial heaters, charging implanted prosthetic devices such as cardiac pacemakers and insulin pumps [1].

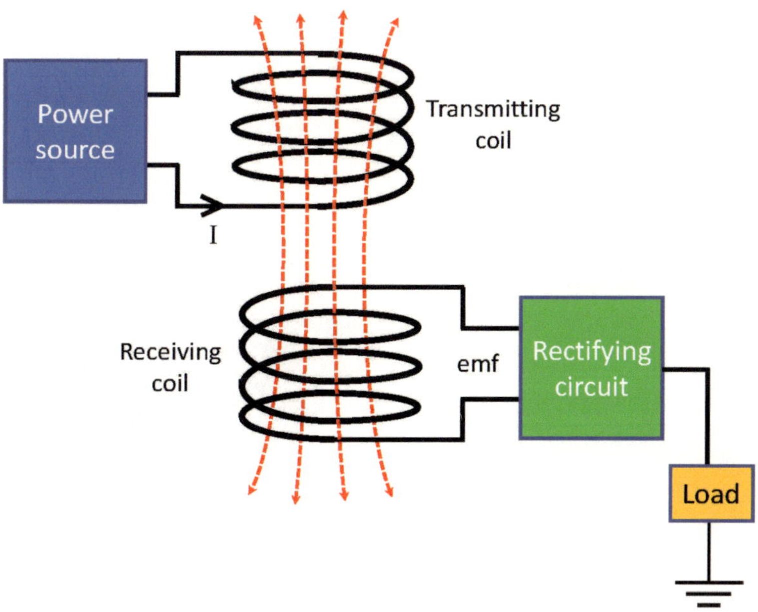

Figure 3.
WPT using inductive coupling scheme.

$$emf = -\frac{d\phi}{dt} \quad (2)$$

WPT system performance can be estimated by the power transfer efficiency (PTE) which depends on the KQ product. K is the coupling coeffect between transmitter and receiver, it is a ratio and varies from 0 to 1 as a maximum value at totally power coupling. Q is the unloaded quality factor of the transmitter's or receiver's coil; Q can be calculated from the coil inductance as in (3), where ω is the angular frequency, L is the coil inductance, and R the loss resistance of the coil. While PTE is calculated from (4) [2]. It is clear that increasing the transfer efficiency needs a high value of the KQ product.

$$Q = \frac{\omega L}{R} \quad (3)$$

$$PTE = \frac{k^2 Q_t Q_r}{\left[1+\sqrt{1+k^2 Q_t Q_r}\right]^2} \times 100\% \quad (4)$$

Numerous studies were introduced in the inductive coupling approach [3–9]. In [10], a multi-layer spiral inductor is proposed for biomedical applications at a frequency of 13.56 MHz which is the license-free industrial, scientific, and medical (ISM) band. It uses a stacked structure to achieve a compact WPT, where the stacked inductors occupying an area of 10 mm × 10 mm with 1 cm separation between transmitter and receiver. The inductance is further increased by stacking the printed spiral inductors on top of each other in such a way that the flow of the current always takes the same direction as shown in **Figure 4**. In [8], a pair of printed spiral coils, as illustrated in **Figure 5**, used in biomedical implanted microelectronic devices to maximize the inductive power transmission efficiency. Zixuan et al. [6] introduced an analysis of alternative-winding coils for getting high-efficiency inductive power for mid-range WPT. Alternative-winding coils structure is demonstrated in **Figure 6**.

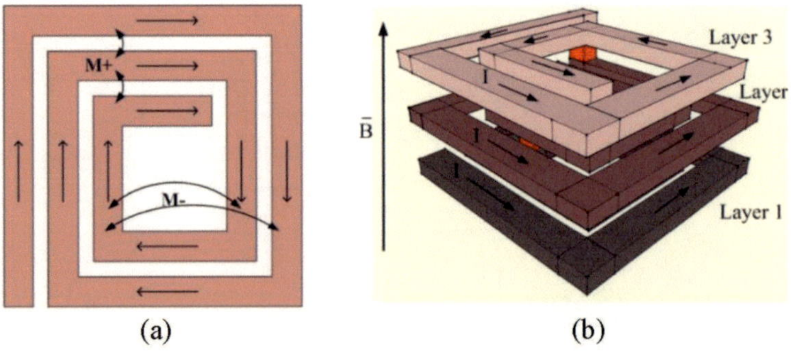

Figure 4.
Multi-layer stacked inductor; (a) top view (b) 3D geometry [10].

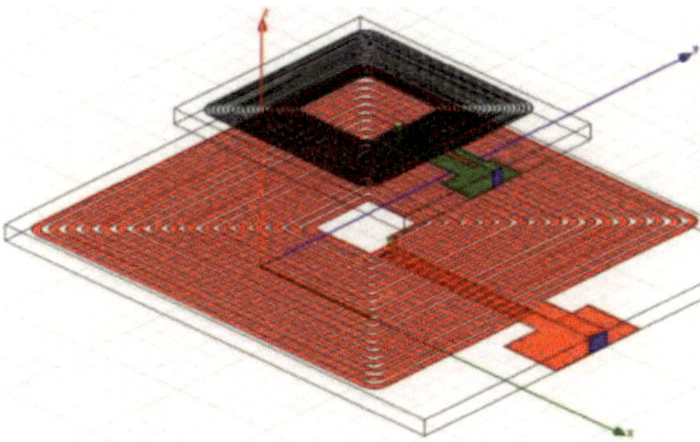

Figure 5.
Design of a pair of printed spiral coils [8].

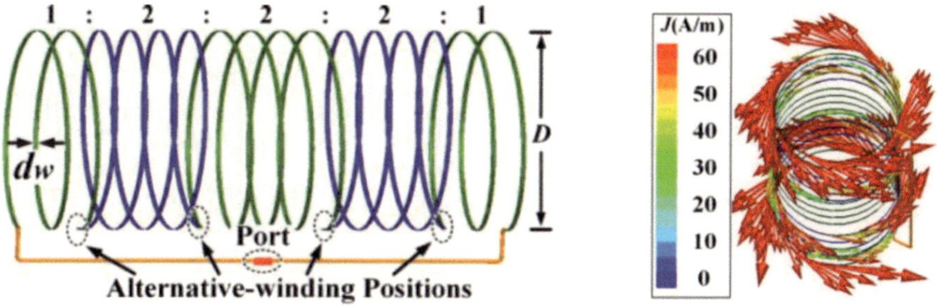

Figure 6.
Alternative-winding coils geometry and its current distribution [6].

3. Resonant inductive coupling WPT

Resonant inductive coupling or magnetic resonance coupling is another form of the WPT technologies in which power is transferred between two tuned resonant circuits, one in the transmitter and the other tuned circuit in the receiver as depicted in **Figure 7**. Each resonant circuit comprises an inductor connected to a capacitor to resonate and couple the transmitted power at their resonance frequency. This resonance is responsible for emphasizing the quality factor (Q-factor) for the resonant circuit. Therefore, the coupling and the power transfer efficiency between the transmitter and receiver increase due to the directly proportional relationship between them. Magnetic resonance coupling scheme is applied in mid-range applications such as charging electric vehicles, charging portable devices, biomedical implants, powering busses, trains, RFID, smartcards.

Several studies have invested the resonant inductive coupling technique for enhancement the power transfer efficiency of WPT systems [11–13]. In [14], we proposed dual open-loop spiral resonators (OLSRs) to improve the magnetic field for WPT system. OLSRs are fed through Metal–Insulator–Metal (MIM) capacitive coupling using a 50 Ohm microstrip transmission line as shown in **Figure 8**. A series resonance model is used to achieve resonant inductive as illustrated in the equivalent circuit model in **Figure 9**. The open-loop spiral resonator (OLSR) includes the series combination between the MIM capacitor and the spiral-loop inductor. Dual OLSRs are used instead of a single OLSR to strengthen the surface current on the spiral resonators. Therefore, it helps to intensify the electromagnetic

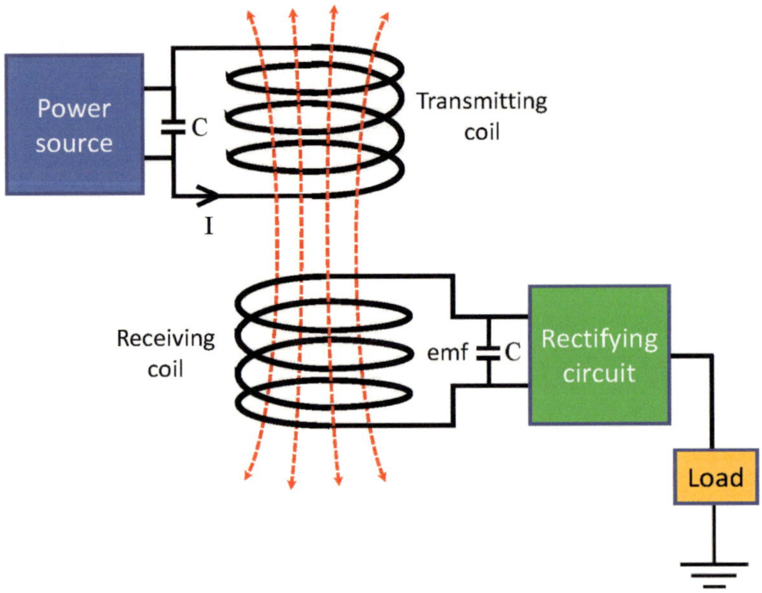

Figure 7.
Resonant inductive coupling WPT structure.

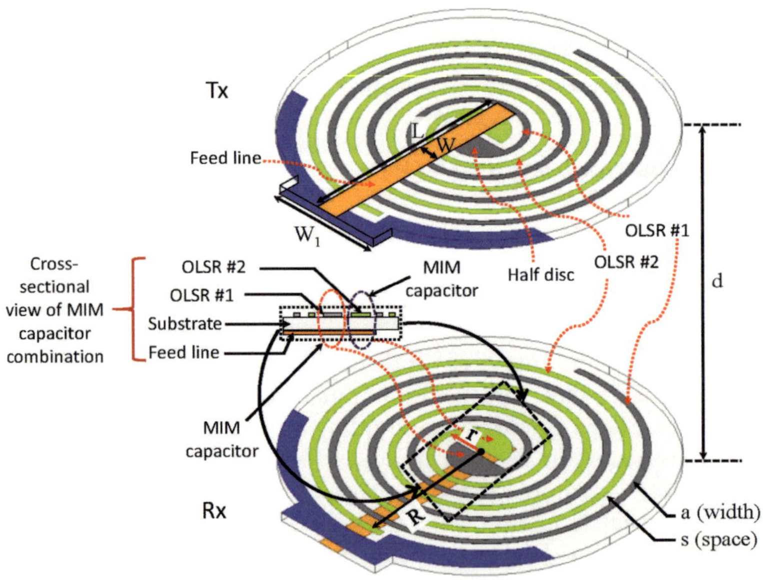

Figure 8.
OLSR WPT geometry [14].

field in order to get a high transmission distance or higher power transfer efficiency. **Figure 10** displays a comparison between the power transfer efficiency for using a single and double OLSR. The results show the improvement in PTE in double OLSR. The OLSRs WPT system operates at 438.5 MHz with a measured PTE of 70.8% at a transmission distance of 31 mm and a design area of 576 mm^2. While PTE for a single OLSR is 56% at 487 MHz at the same transmission distance.

A printed spiral coil with a planar interdigital capacitor is proposed in [15] as shown in **Figure 11**. It studies the misalignment issues between transmitter and

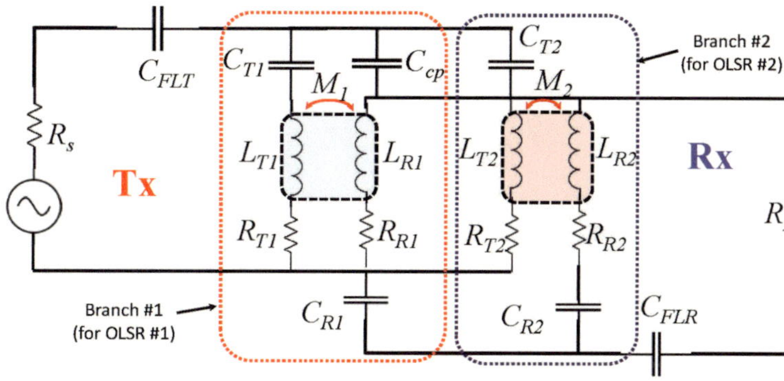

Figure 9.
Equivalent circuit model for OLSR [14].

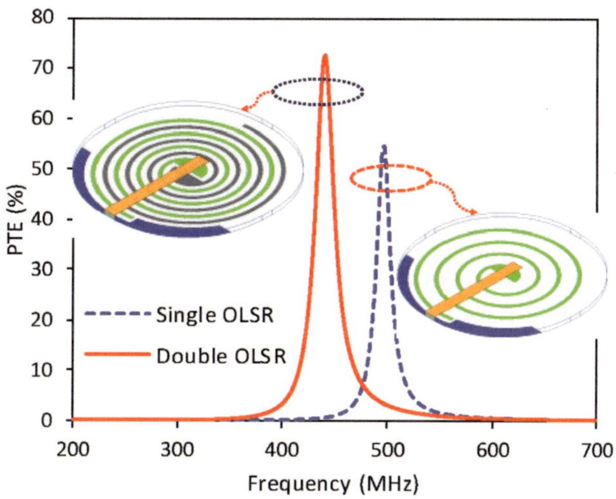

Figure 10.
PTE versus frequency of a single and double OLSR [14].

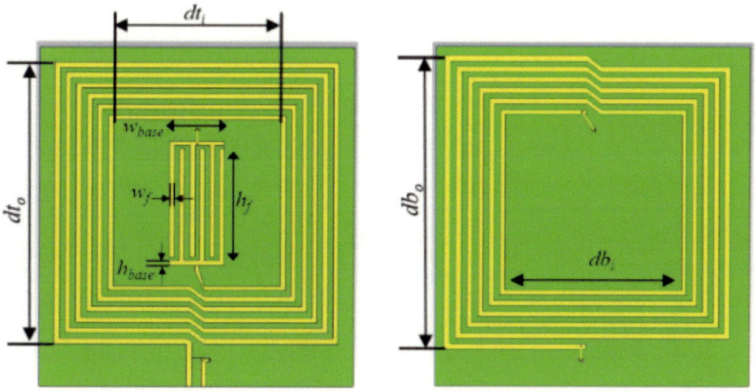

Figure 11.
Geometry of a printed spiral coil with planar interdigital capacitor [15].

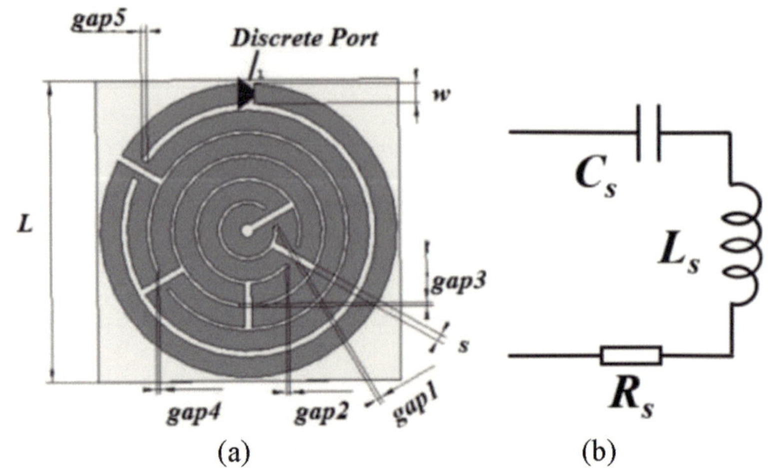

Figure 12.
(a) Conformal split-ring loop self-resonator, (b) equivalent circuit [16].

receiver. Under a perfect alignment, WPT offers a maximum measured transfer efficiency of 71.84%. This research uses the integration between the interdigital capacitor and the spiral coil to get a magnetic resonant resonator with high immunity for the misalignment instances. Wang *et al.* [16] proposed a conformal split-ring loop self-resonator which has a self-resonant frequency and its equivalent circuit is a series resonant circuit composed of an inductor-capacitor series connection as displayed in **Figure 12**. This resonator introduces a high transfer efficiency of 87.9% at a transfer distance of 22 mm. A resonant inductive link for

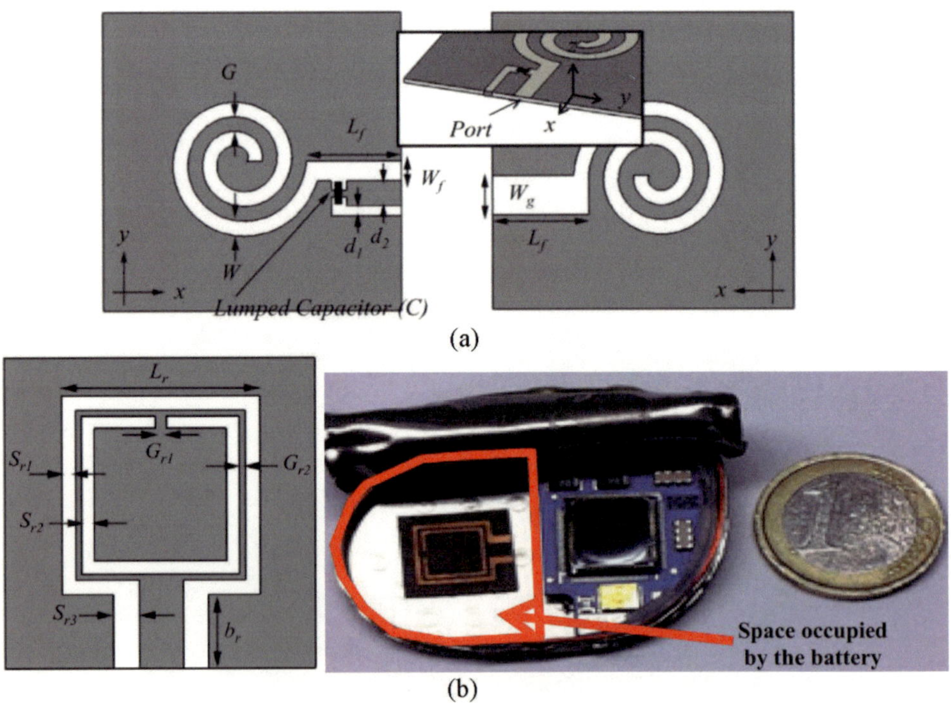

Figure 13.
Spiral coil integrated with lumped capacitor design (a) Transmitter's resonator, (b) Reciever's resonator [17].

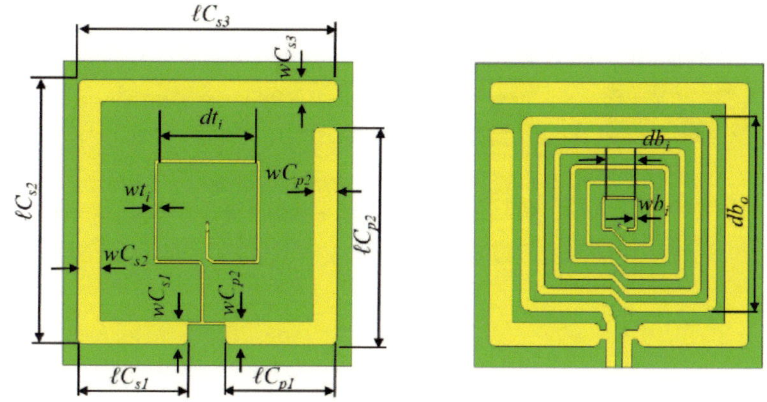

Figure 14.
Capacitive compensated plates design [18].

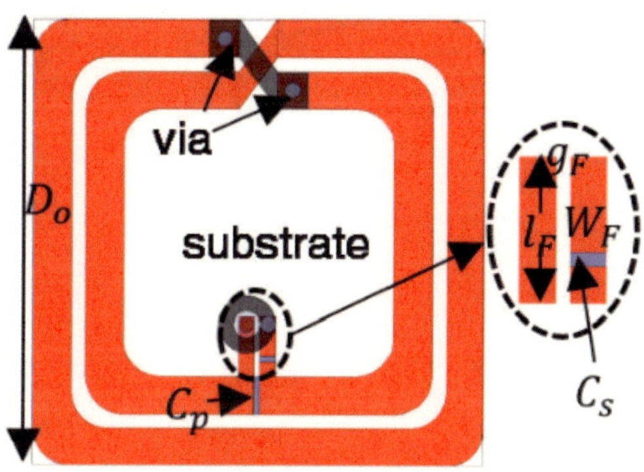

Figure 15.
Planar view of the transmitter/Reciever [19].

powering pacemakers was presented in [17]. The transmitting resonator consists of two spirals printed on the top and bottom face of the Arlon substrate as illustrated in **Figure 13**. A surface-mounted capacitor is inserted in a shunt with the printed spiral to tune the resonance frequency at the desired value. On the other hand, the receiving resonator is a square split-ring resonator. Series–parallel capacitive plates are employed with a printed spiral resonator [18] to get satisfactory tolerance toward angular and lateral displacement. **Figure 14** shows capacitive compensated plates, C-shaped and mirrored L-shaped capacitive plates are formed on the top and bottom layer of the substrate. **Figure 15** presents an asymmetric resonant inductive coupled WPT system [19]. This system has a measured power transfer efficiency of 75% at a transmission distance of 38 mm.

4. Strongly coupled magnetic resonance WPT

Strongly coupled magnetic resonance refers to inserting intermediate resonators with a high-quality factor (Q) in the transmission path between transmitter and receiver as revealed in **Figure 16**, these intermediate resonators are used to emphasize the transferred magnetic power. This technology is categorized as mid-range WPT. In 2007, a group of researchers at the Massachusetts Institute of Technology proposed an experiment using a strongly coupled magnetic resonance technique [20].

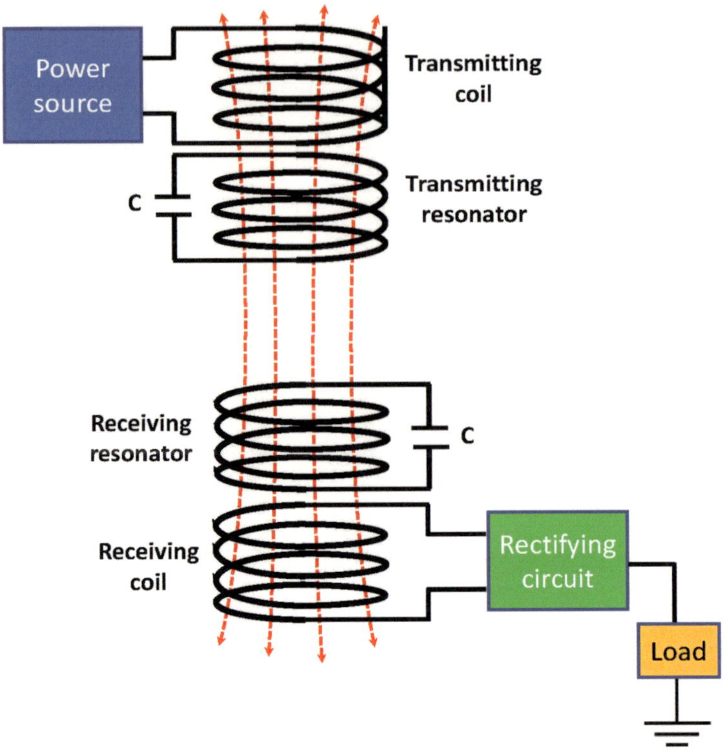

Figure 16.
Strongly coupled magnetic resonance WPT.

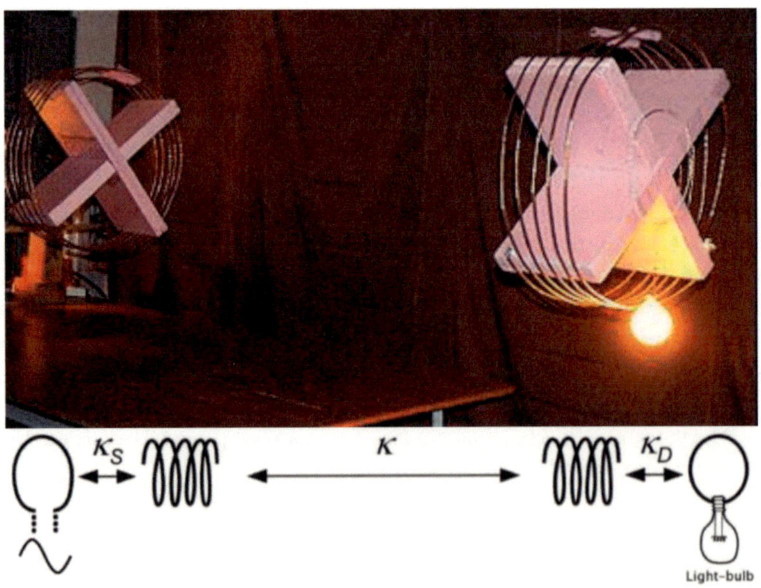

Figure 17.
Setup of MIT researchers group experiment [20].

They effectively powered a light bulb wirelessly using a power source located 2 m away from the light bulb. They obtained a power transfer efficiency of about 40%. The experiment is demonstrated in **Figure 17**, the intermediate resonators are self-resonant.

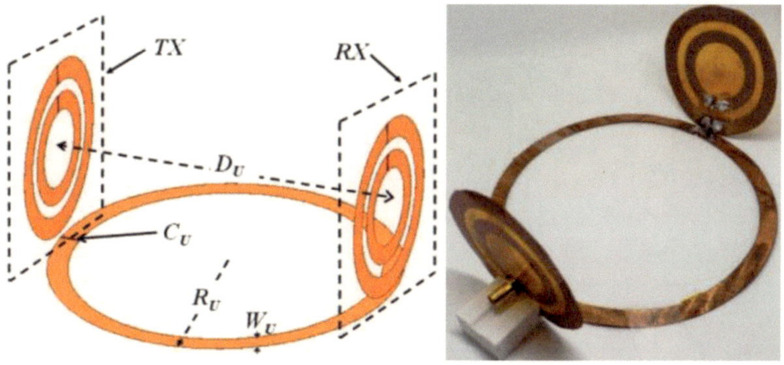

Figure 18.
Conformal strongly coupled magnetic resonance system [26].

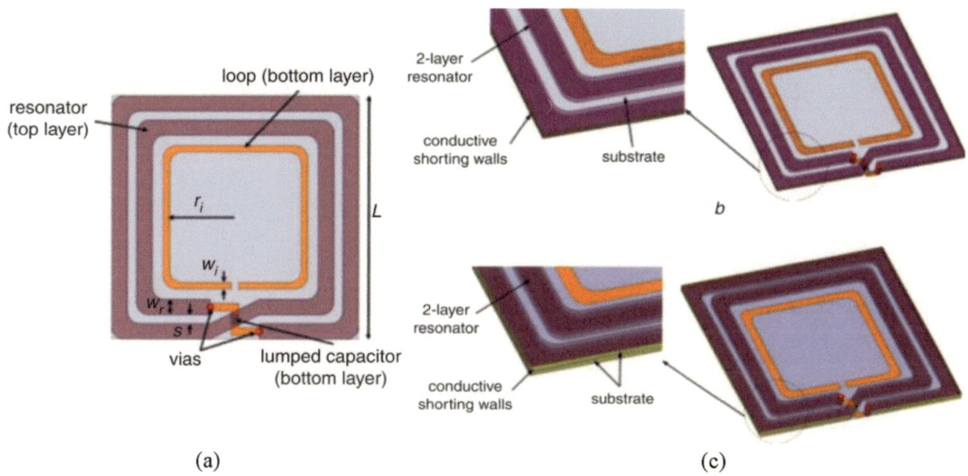

Figure 19.
(a) Geometry of a printed spiral coil, (b) two layers using conductive shorting wall, and (c) three layers using conductive shorting wall [23].

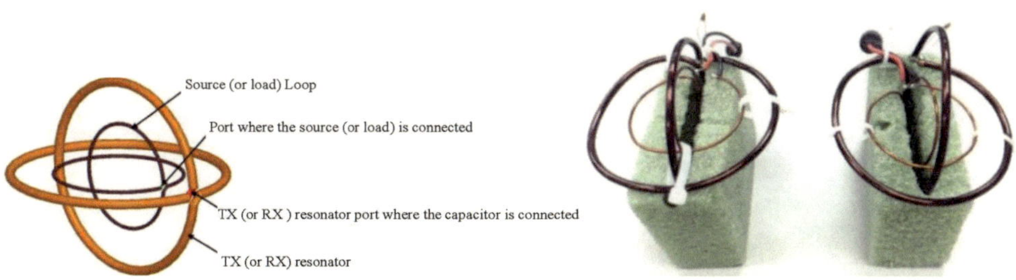

Figure 20.
3-D strongly coupled magnetic resonance WPT [22].

Recently, several authors [21–27], have utilized from the strongly coupled magnetic resonance scheme to enhance the transmission properties of WPT systems. Barreto et al. [26] proposed a conformal strongly coupled magnetic resonance system for range extension by using U-loop as an intermediate resonator as shown in **Figure 18**. It provides a high transfer efficiency reach 70% at a transfer distance equal to the diameter of the U-loop (48 cm). Also, this WPT system can maintain efficiencies greater than 60% regardless of the angular position of the receiver around the U-loop. A multilayer

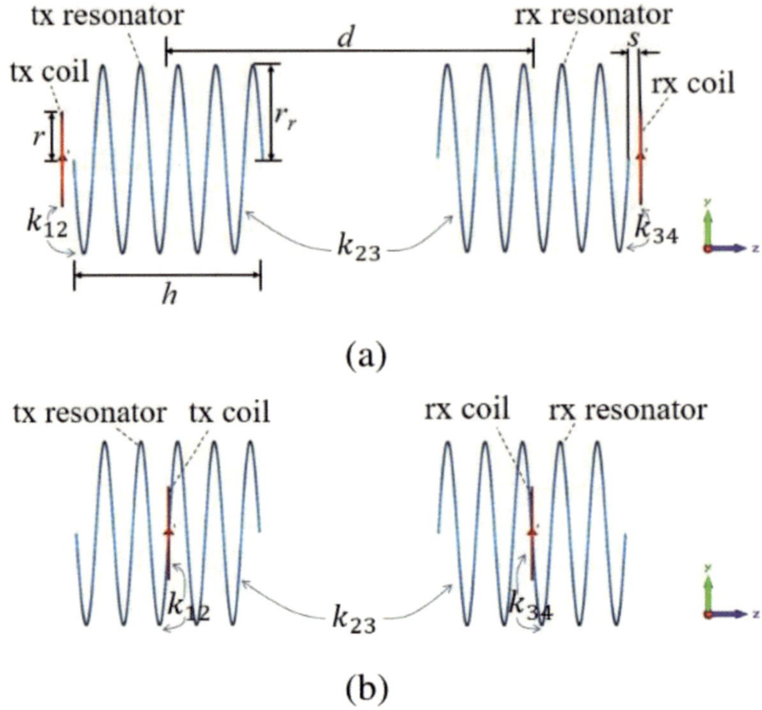

Figure 21.
(a) Conventional four-coil system with the transmitter/receiver coils outside the resonators. (b) Wideband four-coil system with the transmitter/receiver coils at the center of resonators [24].

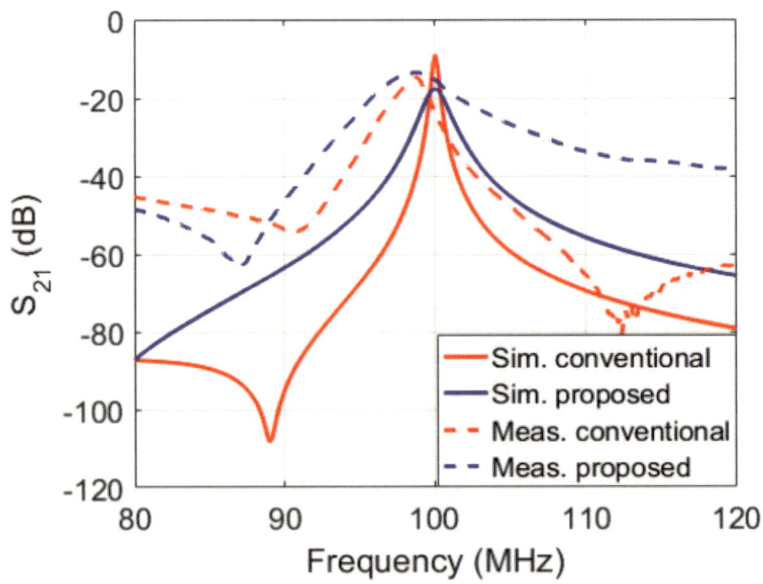

Figure 22.
S_{21} versus frequency [24].

resonator is discussed in [23], where extra layers of printed spiral coils are inserted in the transmitter/receiver resonators to enhance the Q factor and power transfer efficiency. Conductive shorting walls are employed for the connection between the

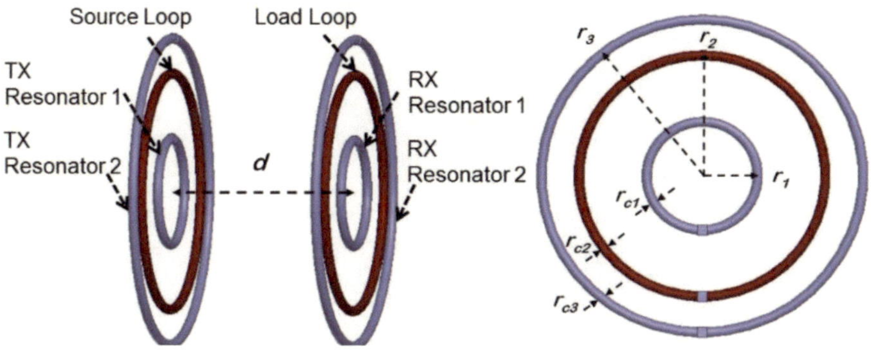

Figure 23.
Configuration for a dual-band conformal strongly magnetic coupling [25].

multilayer resonators as illustrated in **Figure 19**. Liu et al. [22] reduced the misalignment sensitivity of strongly coupled WPT systems by applying two orthogonal coils together in a 3-D model instead of using planar coils as shown in **Figure 20**.

Using strongly coupled magnetic resonance WPT systems leads to getting a high quality factor (Q). Nevertheless, this also results in limiting the system bandwidth. Therefore, Zhou *et al.* proposed a wideband strongly coupled magnetic resonance WPT system [24] to overcome the shifting problems of the resonance frequency that occurs in some practical applications, this, in turn, alleviates the decline in the efficiency caused by this shift in the resonant frequency. **Figure 21** shows the proposed technique, the transmitter and receiver coils are fixed at the center of their corresponding intermediate resonators. In this manner, the leakage of magnetic flux can be mitigated, and the bandwidth is broadened as shown in **Figure 22**. Broadband and multi-band WPT system using conformal strongly coupled magnetic resonance technique is introduced in [25]. A multi-band can be obtained by using multiple pairs of loop resonators with various dimensions to resonate at different frequencies, for example, in **Figure 23**, the source loop and load loop are placed between two resonators (resonator 1 and resonator 2). Each resonator resonates at a different resonance frequency to give a dual-band WPT. The broadband operation can also be achieved by merging between the resonance frequencies, this can be obtained using different values of the capacitance of the loop resonators or use resonators with size near each other. Many designs for multi-band and wideband WPT systems are proposed in [28–35].

5. WPT utilizing meta-surface structures

Metasurface structures are also used to boost the PTE by confining the magnetic field in a narrow channel between transmitter and receiver by combing the evanescent waves from the Transmitter and redirect them into receiver direction due to the negative relative permeability characteristics of some kinds of the metamaterial surfaces. Metamaterials are artificial periodic structures that have negative reflective index characteristics. Metamaterials are classified into three types depending on the polarity of the relative permeability and relative permittivity of the structure: double negative (DNG), ε negative (ENG), and μ negative (MNG), as shown in **Figure 24**.

The inductive, resonance inductive, and strongly coupled magnetic resonance WPT systems rely on the magnetic field coupling between the transmitter and receiver. Thus, the MNG metamaterial category is used with WPT. When the magnetic field travels from the transmitter coil and incident on a metamaterial with MNG, the

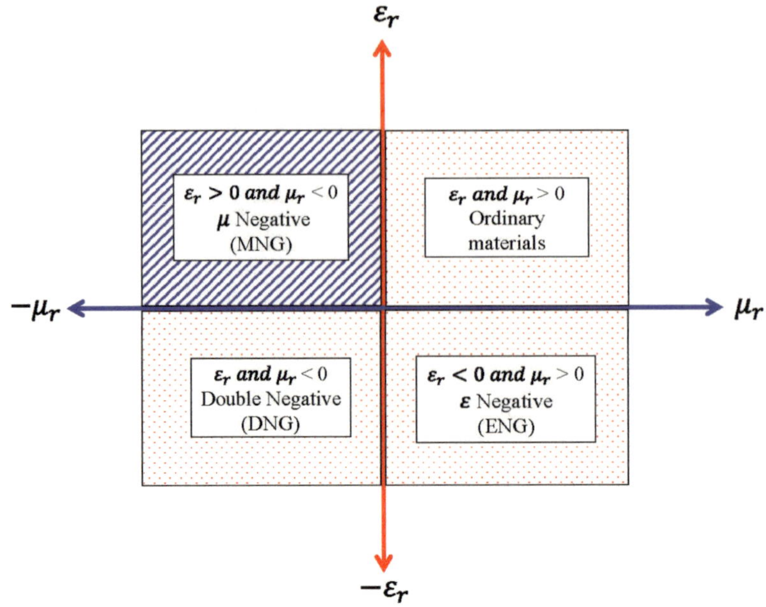

Figure 24.
Metamaterials categories.

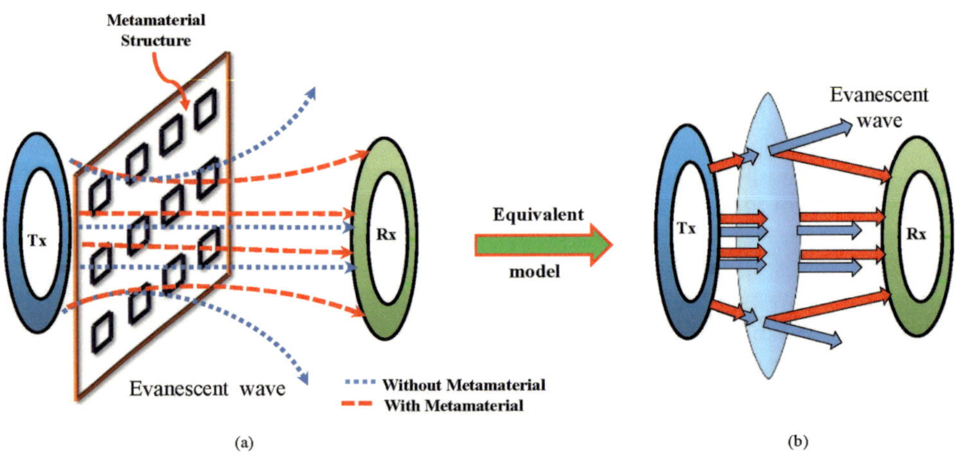

Figure 25.
(a) Metamaterial-based WPT system. (b) equivalent circuit model of applying metamaterial structures with WPT.

outgoing magnetic fields are bent back toward the receiver coil, this increases the field strength between the two coils as revealed in **Figure 25**. Thus, the efficiency is enhanced, and the EMF leakage is reduced due to applying these metamaterial surfaces in the path between the transmitter and receiver coils. **Table 1** summarizes different metamaterial structures that are used in WPT systems [36–40].

6. Capacitive coupling WPT

Capacitive coupling is a kind of coupling that depends on the electric field coupling between two plates, so it is also named electric coupling. Capacitive coupling

Reference	Geometry	Notes
[36]	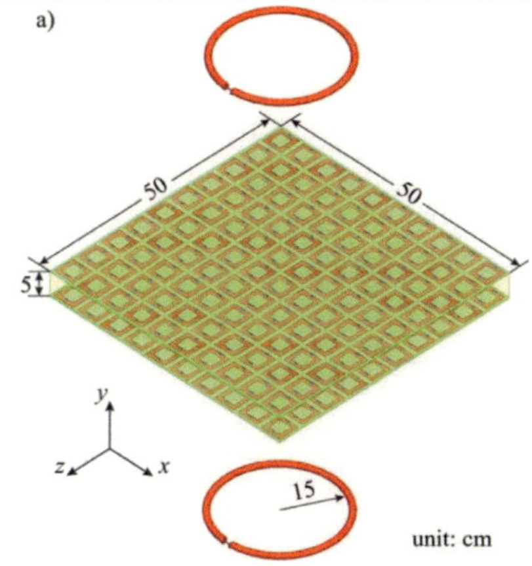	• A thin metamaterial slabs • Study of three types of metamaterials is presented: the double negative material (DNG), the isotropic μ-negative material (MNG), and indefinite material (IM).
[37]	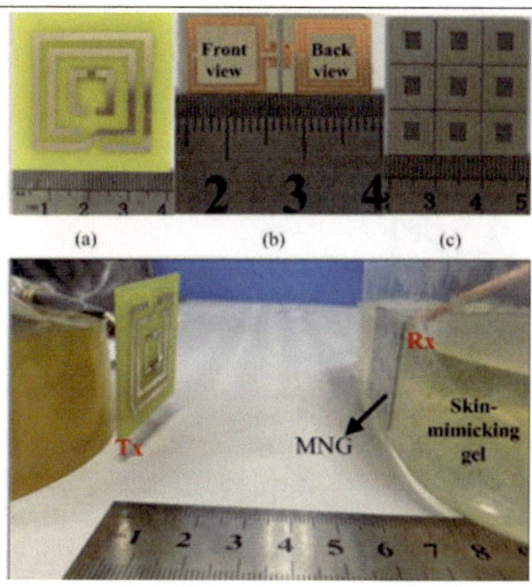	• An efficient wireless power transfer system integrating with negative permeability (MNG) metasurface is proposed for biological applications. • By using metasurface structure, a coupling enhancement of 15.7 dB is obtained.
[38]	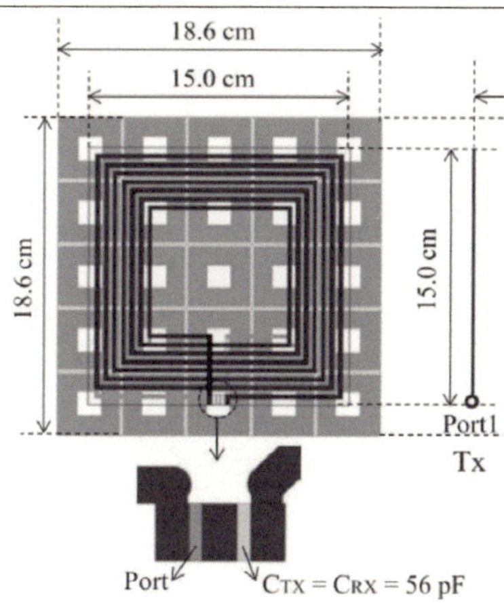	• Metamaterials using a dual-layer printed circuit board (PCB) with a high dielectric constant substrate is proposed for enhancing system efficiency and reduced emf leakage in WPT systems • 44.2% improvement in the PTE and 3.49 dBm reduction in the electromagnetic field leakage at 6.78 MHz and separation distance of 20 cm is obtained.
[39]	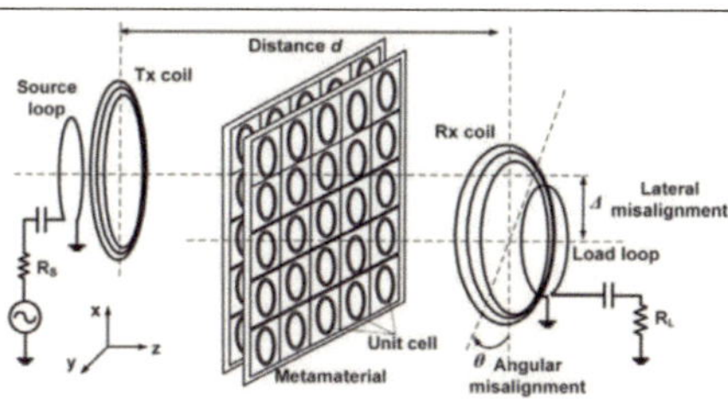	• A study of using metamaterial structure to compensate the degradation in the power transfer efficiency due to the misalignment issues between the Tx and Rx.

Reference	Geometry	Notes
[40]	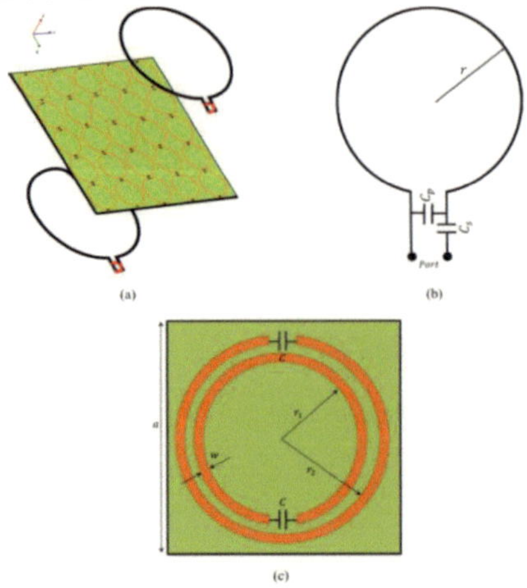	• A closed-form and analytical expressions are obtained for efficiency improvement with metasurface.

Table 1.
Different metamaterial structures used in WPT systems.

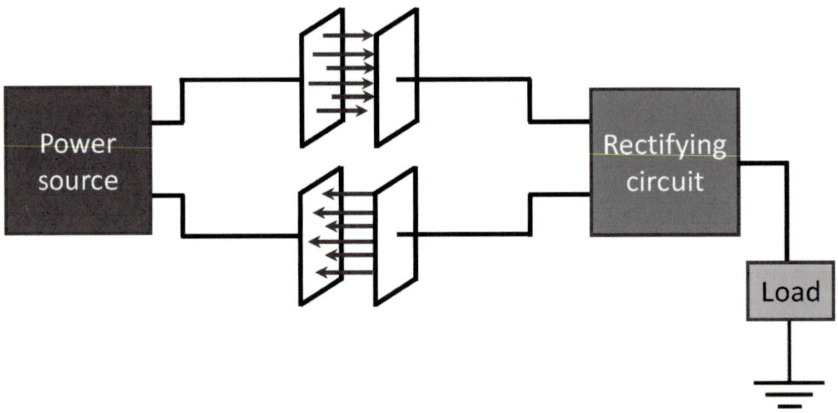

Figure 26.
capacitive wireless power systems.

acts as a capacitor where its metal plates one is in the transmitter and the other in the receiver and the medium in between represents the dielectric. The power can transfer between the two plates in form of a displacement current. **Figure 26** shows the WPT system for the capacitive coupling technique. As a result of electric field interacts with many different materials as well as capacitive coupling method needs very high voltages. Hence, capacitive coupling has only a few practical applications. Capacitive coupling has some special privileges over inductive coupling. The magnetic field is largely confined between the capacitor plates, reducing interference, and higher immunity for the misalignment issues between the transmitter and receiver. Therefore, capacitive coupling can be used in charging portable devices, smartcards, and transferring power between the layers of a substrate in RF integrated circuits. **Figure 27** illustrates an experiment for capacitive coupling that is executed by Nikola Tesla in 1891 [42]. He performed this experiment before his induction WPT demonstration.

In [43], a high-frequency capacitive coupling WPT using dielectric glass layers is introduced to reduce the coupling impedance and increase the coupling

Figure 27.
Tesla demonstrating wireless power transmission using capacitive coupling, New York, in 1891 [41].

capacitance. Thus, it transfers power easily with high efficiency. Regensburger *et al.* introduced a high-performance capacitive WPT system for electric vehicles charging by using interleaved-foil coupled inductors [44]. This system used a kilowatt-scale large air-gap to achieve high power transfer density and high transfer efficiency at the operating frequency (13.56 MHz). Interleaved-foil air-core inductors provide a better quality factor; this makes them are useful at kilowatt-scale power at high frequencies. In [45], multi-loop control that is used to regulate the power transfer in capacitive wireless systems by applying variable matching networks is discussed. An adaptive multi-loop controller combines continuous frequency tracking and matching networks tuning to regulate a current/power to the receiving side at the optimal power transfer conditions. In [46–50], hybrid structures that combine inductive coupling and capacitive coupling WPT in the same system were proposed.

7. Microwave power transfer (MPT)

Microwave power transmission refers to far-field directive powering, where the power transmission occurs in the far-field using a well-defined directional transmitter. Microwave power transmission depends on the propagation of electromagnetic radiative fields where it is preferred in long-range WPT applications. This sort of WPT is useful for space-based solar power satellites (SPS) applications or with intentional powering such as using a dedicating source with a well-known direction to power a network of wireless sensors, each sensor has its built-in rectenna. One of the first applicable trails of MPT was conducted by William Brown et al. in 1965 by powering an aircraft using a MPT at an altitude of fifty feet for ten continuous hours [51].

There are many challenges regarding RF-to-DC power conversion efficiency, matching circuit design, the dependence of the DC output voltage as well as the conversion efficiency on the input power, load impedance, and operating frequency. In order to solve these issues, many rectennas have been introduced [52, 53]. Several single frequency band rectennas were used for energy harvesting [54, 55], and dual and multiband rectennas were discussed in [56–58]. In [59, 60] we proposed a dual-band rectenna using voltage doubler rectifier and four-section matching network. An enhanced-gain antenna with Defected Reflector Structure (DRS) is integrated with the rectifying circuit for increasing the rectenna capability for scavenging. A voltage doubler circuit is used for the rectification. Moreover, a four-section

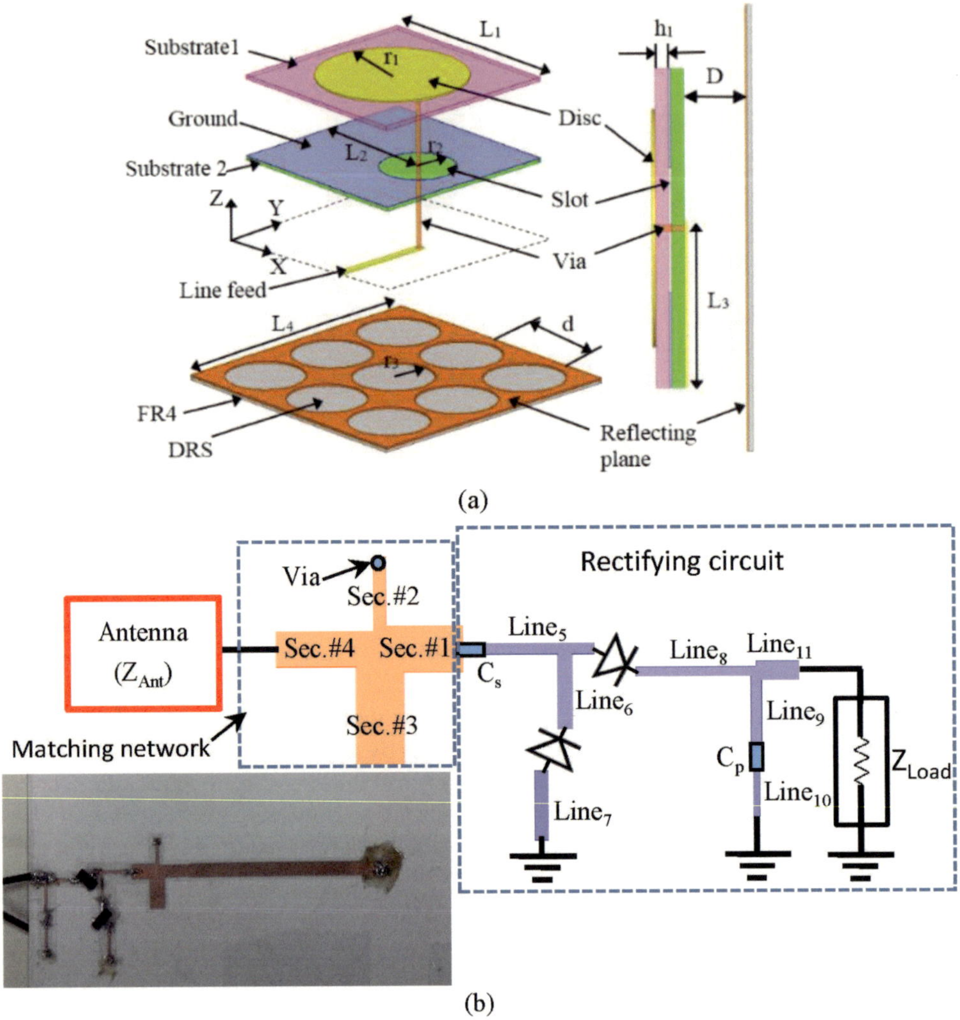

Figure 28.
Dual-band rectenna using four-section matching network, (a) high-gain received antenna, and (b) integration between the receiving antenna and the rectifying circuit [59, 60].

matching network is employed for the matching between the antenna and the rectifier circuit. This matching scheme is used to match between a complex and frequency dependent rectifier input impedance and a real impedance of the antenna (Z_{Ant}) by using different sections (Sec.#1, Sec.#2, Sec.#3, and Sec.#4) as shown in **Figure 28**.

Also in 2020 [61], we proposed a dual-band rectenna for low power applications. The rectenna is comprised of a co-planar (cpw) rectifier integrated with a rectangular split ring antenna loaded by a meandered strip line. A single diode series connection topology is used to miniaturize the losses at low input power operation. For maximum power transfer between the antenna and the rectifying circuit, the matching circuit that consists of a spiral coil in addition to two short circuit stubs is used as shown in **Figure 29**. The proposed rectenna operates at low input power with relatively high measured RF-DC conversion efficiency up to 74% at an input power of −6.5 dBm at the first resonant frequency f_1 = 700 MHz and 70% at −4.5 dBm at the second operating frequency f_2 = 1.4GHz with a resistive load of 1.9 K.

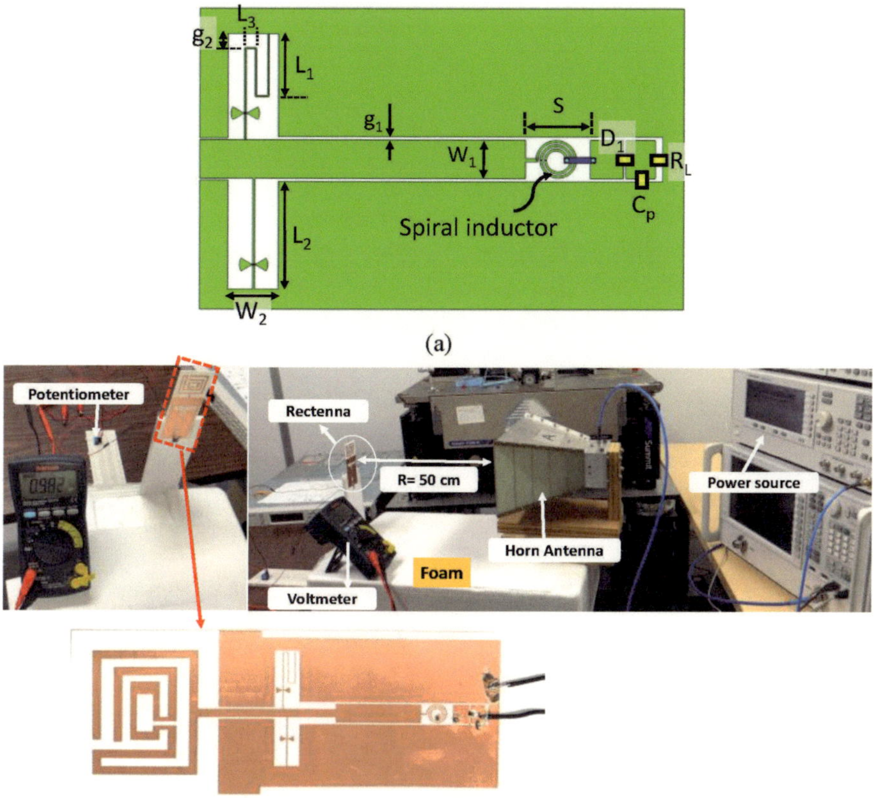

(a)

(b)

Figure 29.
Low power rectenna, (a) rectifier geometry, and (b) measurement setup [61].

8. Conclusion

This chapter presents a study of wireless power transfer technologies. A survey of employing several techniques such as inductive coupling, **resonant inductive coupling,** strongly coupled magnetic resonance, and capacitive coupling for increasing the power transfer efficiency for WPT systems. Metasurface-based WPT systems are also discussed. Many recently published WPT designs are listed with a highlight for the used techniques. Microwave Power Transfer (MPT) also introduced, and two rectenna designs are described.

Author details

Mohamed Aboualalaa[1]*, Hala Elsadek[1] and Ramesh K. Pokharel[2]

1 Electronics Research Institute, Cairo, Egypt

2 Kyushu University, Fukuoka, Japan

*Address all correspondence to: mohamed.ali@ejust.edu.eg

References

[1] "https://en.wikipedia.org/wiki/Wireless_power_transfer ."

[2] T. Ohira, "The kQ Product as Viewed by an Analog Circuit Engineer," *IEEE Circuits Syst. Mag.*, vol. 17, no. 1, pp. 27-32, 2017, doi: 10.1109/MCAS.2016.2642698.

[3] A. Ibrahim and M. Kiani, "A Figure-of-Merit for Design and Optimization of Inductive Power Transmission Links for Millimeter-Sized Biomedical Implants," *IEEE Trans. Biomed. Circuits Syst.*, vol. 10, no. 6, pp. 1100-1111, 2016, doi: 10.1109/TBCAS.2016.2515541.

[4] J. H. Kim *et al.*, "Development of 1-MW Inductive Power Transfer System for a High-Speed Train," *IEEE Trans. Ind. Electron.*, vol. 62, no. 10, pp. 6242-6250, 2015, doi: 10.1109/TIE.2015.2417122.

[5] B. H. Choi, V. X. Thai, E. S. Lee, J. H. Kim, and C. T. Rim, "Dipole-Coil-Based Wide-Range Inductive Power Transfer Systems for Wireless Sensors," *IEEE Trans. Ind. Electron.*, vol. 63, no. 5, pp. 3158-3167, 2016, doi: 10.1109/TIE.2016.2517061.

[6] Z. Yi, M. Li, B. Muneer, and Q. Zhu, "High-efficiency mid-range inductive power transfer employing alternative-winding coils," *IEEE Trans. Power Electron.*, vol. 34, no. 7, pp. 6706-6721, 2019, doi: 10.1109/TPEL.2018.2872047.

[7] R. Matias, B. Cunha, and R. Martins, "Modeling inductive coupling for wireless power transfer to integrated circuits," *2013 IEEE Wirel. Power Transf. WPT 2013*, vol. 2, no. 1, pp. 198-201, 2013, doi: 10.1109/WPT.2013.6556917.

[8] U. M. Jow and M. Ghovanloo, "Design and optimization of printed spiral coils for efficient inductive power transmission," *Proc. IEEE Int. Conf. Electron. Circuits, Syst.*, vol. 1, no. 3, pp. 70-73, 2007, doi: 10.1109/ICECS.2007.4510933.

[9] M. Kiani, U. M. Jow, and M. Ghovanloo, "Design and optimization of a 3-coil inductive link for efficient wireless power transmission," *IEEE Trans. Biomed. Circuits Syst.*, vol. 5, no. 6, pp. 579-591, 2011, doi: 10.1109/TBCAS.2011.2158431.

[10] A. B. Islam, S. K. Islam, and F. S. Tulip, "Design and Optimization of Printed Circuit Board Inductors for Wireless Power Transfer System," *Circuits Syst.*, vol. 04, no. 02, pp. 237-244, 2013, doi: 10.4236/cs.2013.42032.

[11] M. Wagih, A. Komolafe, and B. Zaghari, "Dual-Receiver Wearable 6.78 MHz Resonant Inductive Wireless Power Transfer Glove Using Embroidered Textile Coils," *IEEE Access*, vol. 8, pp. 24630-24642, 2020, doi: 10.1109/ACCESS.2020.2971086.

[12] X. Mou, D. T. Gladwin, R. Zhao, and H. Sun, "Survey on magnetic resonant coupling wireless power transfer technology for electric vehicle charging," *IET Power Electron.*, vol. 12, no. 12, pp. 3005-3020, 2019, doi: 10.1049/iet-pel.2019.0529.

[13] Z. Dai, Z. Fang, H. Huang, Y. He, and J. Wang, "Selective Omnidirectional Magnetic Resonant Coupling Wireless Power Transfer With Multiple-Receiver System," *IEEE Access*, vol. 6, pp. 19287-19294, 2018, doi: 10.1109/ACCESS.2018.2809797.

[14] M. Aboualalaa, I. Mansour, A. Barakat, K. Yoshitomi, and R. K. Pokharel, "Improvement of Magnetic Field for Near-Field WPT System Using Two Concentric Open-Loop Spiral Resonators," *IEEE Microw. Wirel. Components Lett.*, no. 2, pp. 1-4, 2020, doi: 10.1109/lmwc.2020.3016136.

[15] L. L. Pon, C. Y. Leow, S. K. A. Rahim, A. A. Eteng, and M. R. Kamarudin, "Printed Spiral Resonator for Displacement-Tolerant Near-Field Wireless Energy Transfer," *IEEE Access*, vol. 7, pp. 172055-172064, 2019, doi: 10.1109/ACCESS.2019.2893805.

[16] J. Wang *et al.*, "A Conformal Split-Ring Loop as a Self-Resonator for Wireless Power Transfer," *IEEE Access*, vol. 8, pp. 911-919, 2020, doi: 10.1109/ACCESS.2019.2918640.

[17] G. Monti, P. Arcuti, and L. Tarricone, "Resonant Inductive Link for Remote Powering of Pacemakers," *IEEE Trans. Microw. Theory Tech.*, vol. 63, no. 11, pp. 3814-3822, 2015, doi: 10.1109/TMTT.2015.2481387.

[18] L. L. Pon, S. K. Abdul Rahim, C. Y. Leow, M. Himdi, and M. Khalily, "Displacement-tolerant printed spiral resonator with capacitive compensated-plates for non-radiative wireless energy transfer," *IEEE Access*, vol. 7, pp. 10037-10044, 2019, doi: 10.1109/ACCESS.2019.2891015.

[19] A. Barakat, S. Hekal, and R. K. Pokharel, "Simple design approach for asymmetric resonant inductive coupled WPT systems using J-inverters," *Asia-Pacific Microw. Conf. Proceedings, APMC*, vol. 0, pp. 1-3, 2016, doi: 10.1109/APMC.2016.7931382.

[20] A. Kurs, A. Karalis, R. Moffatt, J. D. Joannopoulos, P. Fisher, and M. Soljačić, "Wireless power transfer via strongly coupled magnetic resonances," *Science (80-.).*, vol. 317, no. 5834, pp. 83-86, 2007.

[21] O. Jonah, S. V. Georgakopoulos, and M. M. Tentzeris, "Wireless power transfer to mobile wearable device via resonance magnetic," *2013 IEEE 14th Annu. Wirel. Microw. Technol. Conf. WAMICON 2013*, pp. 1-3, 2013, doi: 10.1109/WAMICON.2013.6572768.

[22] D. Liu, H. Hu, and S. V. Georgakopoulos, "Misalignment sensitivity of strongly coupled wireless power transfer systems," *IEEE Trans. Power Electron.*, vol. 32, no. 7, pp. 5509-5519, 2017, doi: 10.1109/TPEL.2016.2605698.

[23] F. Jolani, Y. Yu, and Z. Chen, "Enhanced planar wireless power transfer using strongly coupled magnetic resonance," *Electron. Lett.*, vol. 51, no. 2, pp. 173-175, 2015, doi: 10.1049/el.2014.4104.

[24] W. Zhou, S. Sandeep, P. Wu, P. Yang, W. Yu, and S. Y. Huang, "A wideband strongly coupled magnetic resonance wireless power transfer system and its circuit analysis," *IEEE Microw. Wirel. Components Lett.*, vol. 28, no. 12, pp. 1152-1154, 2018, doi: 10.1109/LMWC.2018.2876767.

[25] H. Hu and S. V. Georgakopoulos, "Multiband and Broadband Wireless Power Transfer Systems Using the Conformal Strongly Coupled Magnetic Resonance Method," *IEEE Trans. Ind. Electron.*, vol. 64, no. 5, pp. 3595-3607, 2017, doi: 10.1109/TIE.2016.2569459.

[26] J. Barreto, A.-S. Kaddour, and S. V. Georgakopoulos, "Conformal Strongly Coupled Magnetic Resonance Systems With Extended Range," *IEEE Open J. Antennas Propag.*, vol. 1, no. May, pp. 264-271, 2020, doi: 10.1109/ojap.2020.2999447.

[27] B. T. Nukala, J. Tsay, D. Y. C. Lie, J. Lopez, and T. Q. Nguyen, "Efficient near-field inductive wireless power transfer for miniature implanted devices using strongly coupled magnetic resonance at 5.8 GHz," *2016 Texas Symp. Wirel. Microw. Circuits Syst. WMCS 2016*, pp. 4-7, 2016, doi: 10.1109/WMCaS.2016.7577481.

[28] M. Wang, C. Zhou, M. Shen, and Y. Shi, "Frequency drift insensitive broadband wireless power transfer

system," *AEU - Int. J. Electron. Commun.*, vol. 117, p. 153121, 2020, doi: https://doi.org/10.1016/j.aeue.2020.153121.

[29] O. Jonah, S. V. Georgakopoulos, and M. M. Tentzeris, "Multi-band wireless power transfer via resonance magnetic," *IEEE Antennas Propag. Soc. AP-S Int. Symp.*, no. 2, pp. 850-851, 2013, doi: 10.1109/APS.2013.6711084.

[30] D. Liu, K. Bao, Y. Shafiq, and S. V. Georgakopoulos, "Simultaneous Wireless Power and Data Transfer Through Broadband CSCMR," *2018 IEEE Antennas Propag. Soc. Int. Symp. Usn. Natl. Radio Sci. Meet. APSURSI 2018 - Proc.*, pp. 2535-2536, 2018, doi: 10.1109/APUSNCURSINRSM.2018.8608244.

[31] M. Dionigi and M. Mongiardo, "A novel resonator for simultaneous Wireless Power Tranfer and Near Field Magnetic Communications," *IEEE MTT-S Int. Microw. Symp. Dig.*, pp. 61-64, 2012, doi: 10.1109/MWSYM.2012.6259383.

[32] A. Fereshtian and J. Ghalibafan, "Impedance matching and efficiency improvement of a dual-band wireless power transfer system using variable inductance and coupling method," *AEU - Int. J. Electron. Commun.*, vol. 116, p. 153085, 2020, doi: 10.1016/j.aeue.2020.153085.

[33] F. Tahar, A. Barakat, R. Saad, K. Yoshitomi, and R. K. Pokharel, "Dual-Band Defected Ground Structures Wireless Power Transfer System With Independent External and Inter-Resonator Coupling," *IEEE Trans. Circuits Syst. II Express Briefs*, vol. 64, no. 12, pp. 1372-1376, 2017, doi: 10.1109/TCSII.2017.2740401.

[34] F. Tahar, S. Chalise, K. Yoshitomi, A. Barakat, and R. K. Pokharel, "Compact Dual-Band Wireless Power Transfer Using Overlapped Single Loop Defected Ground Structure," *2018 IEEE Wirel. Power Transf. Conf. WPTC 2018*, vol. 1, pp. 1-4, 2018, doi: 10.1109/WPT.2018.8639281.

[35] A. Barakat, S. Alshhawy, K. Yoshitomi, and R. K. Pokharel, "Triple-Band Near-Field Wireless Power Transfer System Using Coupled Defected Ground Structure Band Stop Filters," *IEEE MTT-S Int. Microw. Symp. Dig.*, vol. 2019-June, pp. 1411-1414, 2019, doi: 10.1109/mwsym.2019.8700853.

[36] Y. Z. E. Leelarasmee, "CONTROLLING THE RESONANCES OFINDEFINITE MATERIALS FORMAXIMIZING EFFICIENCY IN WIRELESSPOWER TRANSFER," *Microw. Opt. Technol. Lett.*, vol. 56, no. 3, pp. 867-875, 2014, doi: 10.1002/mop.

[37] L. Li, H. Liu, H. Zhang, and W. Xue, "Efficient Wireless Power Transfer System Integrating with Metasurface for Biological Applications," *IEEE Trans. Ind. Electron.*, vol. 65, no. 4, pp. 3230-3239, 2018, doi: 10.1109/TIE.2017.2756580.

[38] Y. Cho *et al.*, "Thin PCB-type metamaterials for improved efficiency and reduced EMF leakage in wireless power transfer systems," *IEEE Trans. Microw. Theory Tech.*, vol. 64, no. 2, pp. 353-364, 2016, doi: 10.1109/TMTT.2015.2514090.

[39] A. L. A. K. Ranaweera, C. A. Moscoso, and J. W. Lee, "Anisotropic metamaterial for efficiency enhancement of mid-range wireless power transfer under coil misalignment," *J. Phys. D. Appl. Phys.*, vol. 48, no. 45, 2015, doi: 10.1088/0022-3727/48/45/455104.

[40] H. Younesiraad and M. Bemani, "Analysis of coupling between magnetic dipoles enhanced by metasurfaces for wireless power transfer efficiency improvement," *Sci. Rep.*, vol. 8, no. 1, pp. 1-11, 2018, doi: 10.1038/s41598-018-33174-8.

[41] J. Dai and D. C. Ludois, "A Survey of Wireless Power Transfer and a Critical Comparison of Inductive and Capacitive Coupling for Small Gap Applications," *IEEE Trans. Power Electron.*, vol. 30, no. 11, pp. 6017-6029, 2015, doi: 10.1109/TPEL.2015.2415253.

[42] N. Tesla, "Experiments with alternate currents of very high frequency and their application to methods of artificial illumination," *Trans. Am. Inst. Electr. Eng.*, vol. 8, no. 1, pp. 266-319, 1891.

[43] K. H. Yi, "High frequency capacitive coupling wireless power transfer using glass dielectric layers," *2016 IEEE Wirel. Power Transf. Conf. WPTC 2016*, pp. 4-6, 2016, doi: 10.1109/WPT.2016.7498857.

[44] B. Regensburger, S. Sinha, A. Kumar, S. Maji, and K. K. Afridi, "High-Performance Multi-MHz Capacitive Wireless Power Transfer System for EV Charging Utilizing Interleaved-Foil Coupled Inductors," *IEEE J. Emerg. Sel. Top. Power Electron.*, vol. 6777, no. c, pp. 1-1, 2020, doi: 10.1109/jestpe.2020.3030757.

[45] E. Abramov and M. M. Peretz, "Multi-Loop Control for Power Transfer Regulation in Capacitive Wireless Systems by Means of Variable Matching Networks," *IEEE J. Emerg. Sel. Top. Power Electron.*, vol. 8, no. 3, pp. 2095-2110, 2020, doi: 10.1109/JESTPE.2019.2935631.

[46] W. Zhou, Y. G. Su, L. Huang, X. D. Qing, and A. P. Hu, "Wireless power transfer across a metal barrier by combined capacitive and inductive coupling," *IEEE Trans. Ind. Electron.*, vol. 66, no. 5, pp. 4031-4041, 2019, doi: 10.1109/TIE.2018.2849991.

[47] F. Lu, H. Zhang, H. Hofmann, and C. C. Mi, "An Inductive and Capacitive Combined Wireless Power Transfer System with LC-Compensated Topology," *IEEE Trans. Power Electron.*, vol. 31, no. 12, pp. 8471-8482, 2016, doi: 10.1109/TPEL.2016.2519903.

[48] B. Minnaert and N. Stevens, "Conjugate image theory for non-symmetric inductive, capacitive and mixed coupling," *WPTC 2017 - Wirel. Power Transf. Conf.*, vol. 1, no. 8, 2017, doi: 10.1109/WPT.2017.7953857.

[49] S. Gladchenko, M. Khalil, C. J. Lobb, F. C. Wellstood, and K. D. Osborn, "Superposition of inductive and capacitive coupling in superconducting LC resonators," *IEEE Trans. Appl. Supercond.*, vol. 21, no. 3 PART 1, pp. 875-878, 2011, doi: 10.1109/TASC.2010.2089774.

[50] F. Lu, H. Zhang, H. Hofmann, and C. C. Mi, "An Inductive and Capacitive Integrated Coupler and Its LCL Compensation Circuit Design for Wireless Power Transfer," *IEEE Trans. Ind. Appl.*, vol. 53, no. 5, pp. 4903-4913, 2017, doi: 10.1109/TIA.2017.2697838.

[51] W. C. Brown, "Experimental airborne microwave supported platform," RAYTHEON CO BURLINGTON MA SPENCER LAB, 1965.

[52] V. Palazzi, M. Del Prete, and M. Fantuzzi, "Scavenging for Energy: A Rectenna Design for Wireless Energy Harvesting in UHF Mobile Telephony Bands," *IEEE Microw. Mag.*, vol. 18, no. 1, pp. 91-99, 2017, doi: 10.1109/MMM.2016.2616189.

[53] H. Sun and W. Geyi, "A New Rectenna Using Beamwidth-Enhanced Antenna Array for RF Power Harvesting Applications," *IEEE Antennas Wirel. Propag. Lett.*, vol. 16, pp. 1451-1454, 2017, doi: 10.1109/LAWP.2016.2642124.

[54] S. Chandravanshi and M. J. Akhtar, "Design of efficient rectifier using IDC and harmonic rejection filter in GSM/CDMA band for RF energy harvesting," *Microw. Opt. Technol. Lett.*, vol. 59, no. 3,

pp. 681-686, Mar. 2017, doi: https://doi.org/10.1002/mop.30365.

[55] M. Zeng, A. S. Andrenko, X. Liu, Z. Li, and H. Tan, "A Compact Fractal Loop Rectenna for RF Energy Harvesting," *IEEE Antennas Wirel. Propag. Lett.*, vol. 16, pp. 2424-2427, 2017, doi: 10.1109/LAWP.2017.2722460.

[56] P. Lu, X. Yang, J. Li, and B. Wang, "A Compact Frequency Reconfigurable Rectenna for 5.2- and 5.8-GHz Wireless Power Transmission," *IEEE Trans. Power Electron.*, vol. 30, no. 11, pp. 6006-6010, 2015, doi: 10.1109/TPEL.2014.2379588.

[57] S. Shen, C. Chiu, and R. D. Murch, "A Dual-Port Triple-Band L-Probe Microstrip Patch Rectenna for Ambient RF Energy Harvesting," *IEEE Antennas Wirel. Propag. Lett.*, vol. 16, pp. 3071-3074, 2017, doi: 10.1109/LAWP.2017.2761397.

[58] P. Lu, X.-S. Yang, J.-L. Li, and B.-Z. Wang, "A dual-frequency quasi-pifa rectenna with a robust voltage doubler for 2.45- and 5.8-GHz wireless power transmission," *Microw. Opt. Technol. Lett.*, vol. 57, no. 2, pp. 319-322, Feb. 2015, doi: https://doi.org/10.1002/mop.28841.

[59] M. Aboualalaa, A. B. Abdel-Rahman, A. Allam, H. Elsadek, and R. K. Pokharel, "Design of a Dual-Band Microstrip Antenna With Enhanced Gain for Energy Harvesting Applications," *IEEE Antennas Wirel. Propag. Lett.*, vol. 16, pp. 1622-1626, 2017, doi: 10.1109/LAWP.2017.2654353.

[60] M. Aboualalaa, I. Mansour, M. Mansour, A. Bedair, A. Allam, M. Abo-Zahhad, H. Elsadek, K. Yoshitomi, and R. K. Pokharel, "Dual-band Rectenna Using Voltage Doubler Rectifier and Four-Section Matching Network," in *2018 IEEE Wireless Power Transfer Conference (WPTC)*, 2018, pp. 1-4, doi: 10.1109/WPT.2018.8639451.

[61] M. Aboualalaa, I. Mansour, A. Bedair, A. Allam, M. Abo-Zahhad, H. Elsadek, and R. K. Pokharel "Dual-band CPW rectenna for low input power energy harvesting applications," *IET Circuits, Devices Syst.*, vol. 14, no. 6, pp. 892-897, 2020, doi: 10.1049/iet-cds.2020.0013.

Chapter 8

A Defected Metasurface for Field-Localizing Wireless Power Transfer

Sarawuth Chaimool, Chawalit Rakluea, Yan Zhao and Prayoot Akkaraekthalin

Abstract

The potential of wireless power transfer (WPT) has attracted considerable interest for various research and commercial applications for home and industry. Two important topics including transfer efficiency and electromotive force (EMF) leakage are concerned with modern WPT systems. This work presents the defected metasurface for localized WPT to prevent the transfer efficiency degraded by tuning the resonance of only one-unit cell at the certain metasurface (MTS). Localization cavities on the metasurface can be formed in a defected metasurface, thus fields can be confined to the region around a small receiver, which enhances the transfer efficiency and reduces leakage of electromagnetic fields. To create a cavity in MTS, a defected unit cell at the receiving coils' positions for enhancing the efficiency will be designed, aiming to confine the magnetic field. Results show that the peak efficiency of 1.9% for the case of the free space is improved to 60% when the proposed defected metasurface is applied, which corresponds to 31.2 times enhancements. Therefore, the defected MTS can control the wave propagation in two-dimensional of WPT system.

Keywords: localization, wireless power transfer, metasurface, magneto inductive wave, reduced EMF

1. Introduction

Nowadays, wireless technologies are important on our life societies. It is not only on emerging wireless communication systems e.g., 5G, WiFi6, but also on wireless power transfer (WPT) systems. A WPT system can deliver the electrical energy from one to another across the air gap without the need for wires or exposed contacts. An example of the most commonly used WPT is the charging systems of mobile phones and electronics gadget devices. Recently, iPhone12 has been released and all models feature wireless charging [1]. Its power is up to 15 W with the most up-to-date *Qi*- standard [2]. However, most of the mobile wireless charging is not truly wireless because a phone needs to touch a charging station. Another application of WPT technology is an electric vehicle (EV) charging station [3–5], which requires high power and without touching the charging station. The required power is up to a few kilowatts. Although the two systems seem different, both share the same technique, which are used in a near-field WPT. The near-field WPT can be categorized into two types: inductive coupling and magnetic resonance coupling [6]. Usually, the inductive coupling technique is lowering efficiency and shorter distance compared to the magnetic resonance method. The magnetic resonance method is based on the resonant coupling with two-same frequency resonant circuits while it is interacting weakly with other off-resonant frequency. Therefore, many studies have been proposed the techniques to improve efficiency and operating distance [7–10]. Moreover, a near-field WPT system is using magnetic field coupling and uses in high power applications, so electromagnetic fields (EMFs) accordingly increase. Hence, leakage EMF emitted from WPT system should follow the feasible guideline of ICNIRP (International Commission on Non-Ionizing Radiation Protection (ICNIRP) [11]. It can be damaging human bodies and surroundings. To meet safety requirements below a certain standard level, most of the previous works have been proposed using ferrite and metallic shields [12–14]. The ferrite-based shielding is moderately effective, but it is very heavy weight and normally used in low-frequency. While the metallic-based shielding is flatter and more lightweight, the reflected wave can degrade the overall efficiency of the WPT system. Both transfer efficiency and EMF leakage are concerned in modern WPT. Unfortunately, each proposed technique can solve only one by one. Recently, the use of electromagnetic metamaterials (MTM)/metasurface (MTS) for WPT systems have been proposed [15–19]. Metamaterials are engineered composites that exhibit unusual electromagnetics properties, which are not found in natural materials. Usually, implemented MTMs/MTSs are always used a uniform, which all unit cell is identical. Unfortunately, the most property of the uniform MTM/MTS is negative permeability, which can enhance only the transfer efficiency with evanescent wave amplification, hence increasing transfer efficiency. Consequently, many research efforts have been improved the transfer efficiency and decreased leakage electromagnetic fields (EMF) simultaneously. Recently, MTM/MTS have been proposed for enhancing the transfer efficiency and decreasing EMF leakage [20–25]. To solve both efficiency and safety problems, the non-uniform [18] and hybrid MTM/MTS techniques [22] have been proposed, however, these are suitable for the size of transmitting and receiving coils are comparable. Another method is using magneto-inductive wave (MIW), which is supported by MTM/MTS composed of inductively coupled electrically small resonators and created by inter-element couplings [26–29]. The MIWs can be constructed in one, two and three dimensions. Usually, however, the efficiency is fluctuation due to the standing MIW, which depends on the receiver position [27]. Moreover, the two-dimension MIWs have more complex dispersion [27, 29].

In practical various applications, a receiving side such as charging portable or implantable devices is usually smaller than a transmitting side. Moreover, users prefer to be able to freely location the receiving devices. When the receiver size is reduced to the transmitter, the efficiency rapidly decreases either with or without the MTM/MTS. This is because power distribution is spread out and unrefined above the charging surface; that is the transfer efficiency is variation when a receiving device position is misaligned, which is deviated from the concentric position with respect to a transmitter [19]. To refine and control the magnetic field by focusing the fields into the small regions, the non-uniform and defected MTSs with modified field localization technique have been proposed [20, 25, 30, 31]. The defected MTS is formed by the cavities on MTS. To create a cavity, a defected unit cell at the small receiving coils' positions can be designed and controlled by tunning the resonant frequency using external capacitor. The defected unit cell can be locally modified for magnetic field control at subwavelength scale. In addition, it does not change the macroscopic parameters including negative permeability. Unlike the previous approach [20, 25, 30], the receiver location is placed at center of the MTS. There are different mechanisms to create the cavity on the MTS. In [30], the cavity is created by Fano-type interference, while the hybridization bandgap is used in [20] and MIW is applied in [25]. Motivated by these observations, we modify the cavity mode concept to create a new defected metasurface (MTS) for enhancing transferred efficiency and reducing the electromotive force (EMF). The MTS is based on a defected unit cell at the desired receiving position formed a two-dimensional cavity with configuration of the conventional WPT system. Besides, a free space case, a conventional uniform MTS and the proposed defected MTS have been studied and compared.

2. System configuration and metasurface characteristics

2.1 System configuration

A common WPT system consists of a transmitting coil and a receiving coil. Firstly, we investigate four configurations of the conventional WPT systems with and without uniform metasurface: (i) large Tx and large Rx coils (LTLR) without metasurface, (ii) large Tx coil and small Rx coil (LTSR) without metasurface, (iii) LTLR with uniform metasurface, and (iv) LTSR with uniform metasurface as shown in **Figure 1(a)** to **Figure 1(d)**, respectively. These four configurations are the reference configurations to compare with the proposed defected metasurface. The large size of coils is compared with the unit cell size of an MTS. For uniform MTS, all unit cells are identical, and its property is the negative effective permeability. The transmitting and receiving coils are used planar four-coils WPT system configuration and the operating frequency is 13.56 MHz band, which is an ISM (Industrial, Scientific and Medical) band. The large transmitting coil composes of the feed/load coil and three-turns planar resonator coil with a trace width of 3 mm and 17 mm. The gap between the adjacent loop of the resonator coil is 5 mm with the largest diameter of 240 mm. The radius of the feed/load coil is 48.5 mm. Due to the requirement of ISM bands; it is vital to fix the resonant frequency of WPT system. To resonate at a desired frequency of 13.56 MHz, a 68 pF chip capacitor is connected in series. The details are shown in **Figure 1(a)**. For the purposes of comparison, two different sizes of the receiving coil are examined. The first size of the receiving coil is identical to the large transmitting coil and the second one is smaller than the transmitting coil about five times, which the largest diameter is 60 mm. It is a two-turn planar coil and loaded with two capacitors at the driving loop coil and resonant loop coil. The detailed design of the two-turn planar coil is similar to our proposed in [32]. To enhance the efficiency and focus the magnetic

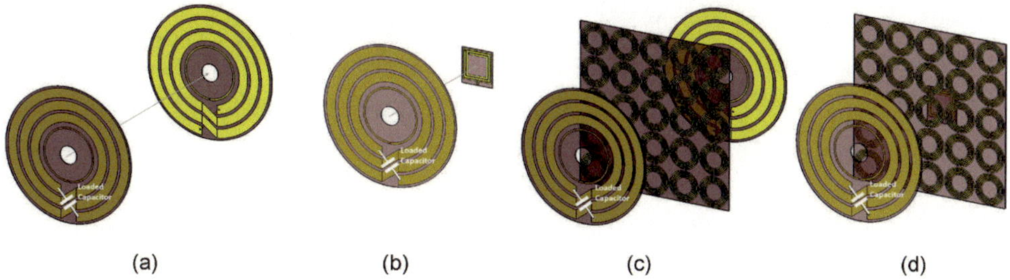

Figure 1.
Configuration of the conventional WPT system. (a) large Tx and large Rx coils (LTLR) without metasurface, (b) large Tx coil and small Rx coil (LTSR) without metasurface, (c) LTLR with uniform metasurface, and (d) LTSR with uniform metasurface.

field, the uniform MTS is placed between the transmitting and receiving coils as shown in **Figure 1(c)** and **Figure 1(d)**. All transmitting coil, receiving coil and unit cell are designed on FR-4 material with a dielectric constant (ε_r) of 4.3 and loss tangent (tan δ) of 0.025.

Then, to study the performance of the proposed defected metasurface on the WPT system, a system configuration is shown in **Figure 2**. It is composed of three parts including a transmitting coil, a defected MTS and a receiving coil. For the LTLR cases, the separation between the transmitting-to-metasurface and receiving-to-metasurface is the equal to 240 mm, so the distance from the transmitting to receiving coils is the total 480 mm as shown in **Figure 2(a)**. As we mentioned previous section, users prefer to able to freely location the receiver. A model of freely multiple receiver locations shows in **Figure 2(b)**. This configuration is not only freely movement the receiver, but also like misalignment between transmitter and receiver when the location is positioned on No. 2 to No. 6, so the efficiency deviates from the center (No. 1). Then, the effects of localization field have been extensively studied when the small receiving coil is placed closer to the defected MTS as shown in **Figure 2(b)**. The distance between the small receiving coil and the defected MTS is 40 mm, whereas the transmitting side is keeping the same. Numbers (No. 1 to No. 6) on unit cells are positioned of the cavities or hotspots formed in the defected MTS. The defected unit cell formed cavity is designed with different resonant frequencies from the uniform MTS. The resonant frequency of the defected unit cell is higher than the uniform one, which can be tuned by using the series chip capacitors. We examine the effects of the positions on the defected MTS using an EM simulator and measurement. The results will show and discuss in the following sections.

2.2 Metasurface characteristics

The metasurface is usually constructed using locally resonant unit cells in the deep subwavelength scale. For realized a compact metasurface and compromise between magnetic field enhancement, an array of 5 × 5-unit cells is chosen in this work. Each unit cell is a single-side planar 5-turn spiral with loaded a capacitor in order to resonate at the desired frequency as shown in **Figure 3(a)**. A loading capacitor of 117 pF is used to resonate at 11.88 MHz. It can be seen in **Figure 3(a)**

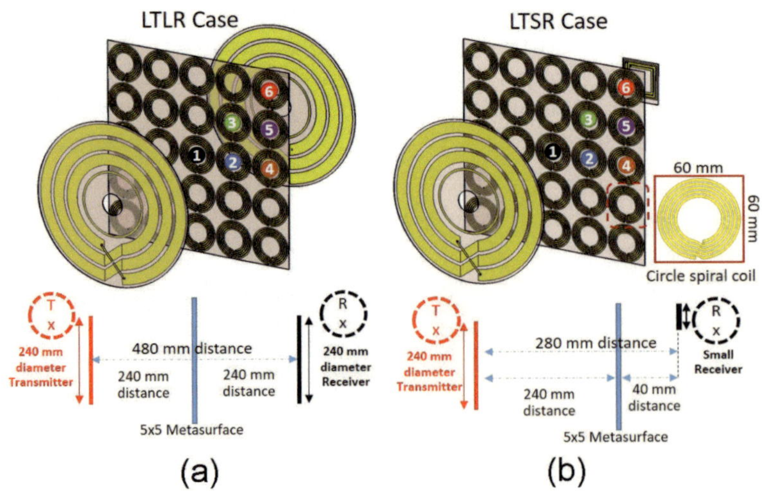

Figure 2.
Configuration of the proposed defected metasurface. (a) LTLR and (b) LTSR. The defected positions are numbering from No. 1 (center) to No. 6.

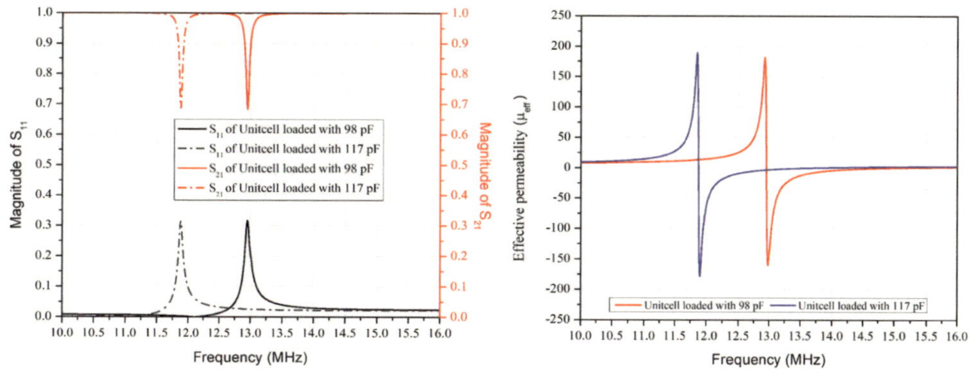

(a) (b)

Figure 3.
Metasurface characteristics (a) resonant frequency and (b) effective permeability.

that the resonant frequencies of 117 pF and 98 pF loadings are 11.88 MHz and 12.95 MHz, respectively. The frequency response of the 98 pF loading is also shown and compared because this configuration is designed for the selective created cavity. The resonant frequency of the cavity is slightly higher than the uniform unit cell. Then, the effective permeability of the proposed MTS can be obtained by using the CST simulation [33] with extracting the S-parameters of the proposed unit-cell as shown in **Figure 3(b)**. Since the WPT system is based on magnetic field coupling when an incident EM wave with a magnetic field is perpendicular to the plane of the MTS, the MTS obeys the frequency dispersive Lorentz model to produce an effective negative permeability. At 13.56 MHz, the effective permeability of both capacitor-loading achieves the negative permeability. To adjust the resonant frequency, it can be changed to the series capacitor [18].

3. Results and discussion

To compare the transfer efficiency (η) of all configurations, we use the magnitude of $|S_{21}|$, which can be easily measured using a vector network analyzer (VNA) in experiments. By using the $|S_{21}|$ and $|S_{11}|$, the function of the transfer efficiency on $\eta = |S_{21}|^2/(1-|S_{11}|^2)$, however, when the network is matching at both ports, the transfer efficiency equals $|S_{21}|^2$. A comparison of simulated $|S_{21}|$ for the four reference cases and the LTLR with the defected MTS (**Figure 2(a)**) is shown in **Figure 4**. Compared with the free space, the magnitude of S_{21} is increased when a uniform or a defected MTS is used. At 13.56 MHz, the uniform MTS case has a maximum power transfer. It can be observed that a case of free-space case (without MTS) and the uniform MTS have only a single peak. The magnitude of S_{21} for the free-space case ($|S_{21}| = 0.24$) is remarkable to lower compared with the case of the uniform MTS ($|S_{21}| = 0.69$). When the defected unit cell is placed in each position, the magnitudes of $|S_{21}|$ are separated into two or more peaks, called *frequency splitting*, which its mechanism is totally different from the two-coil system as a function of separation distance between coils. In a two-coil system, generally, when the distance between transmitting and receiving coils is closer and smaller than a threshold value, it creates two frequency splitting due to the magnetic over-couplings. Hence, many research efforts have been developed the system performance against frequency splitting using optimizing and compensation methods such as non-identical resonant coil [34]. The frequency splitting phenomenon of the defected MTS occurs when non-identical resonant unit cells. It can be explained in terms of Fano

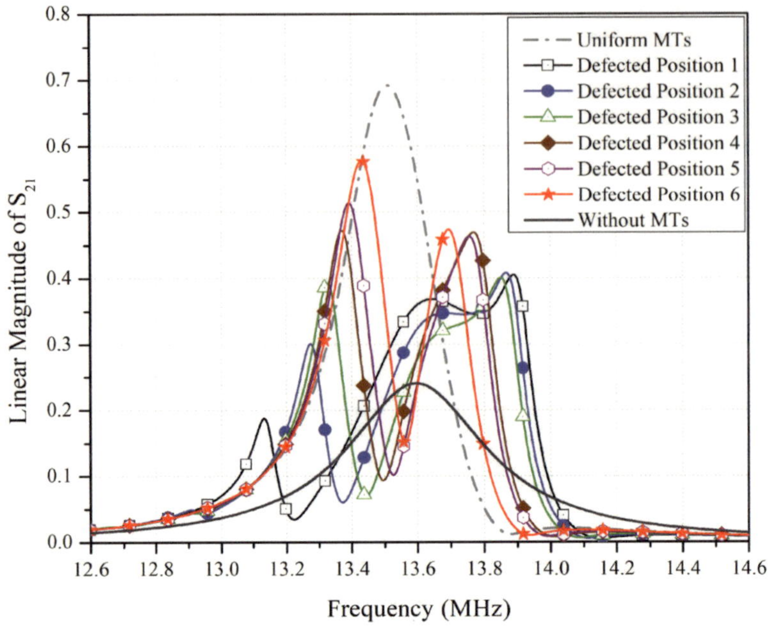

Figure 4.
Comparison of simulated |S_{21}| for LTLR depending on the frequency.

interference [30, 35], which is much more sensitive to the defected position in its periodicity. It is observed that there are two peaks and the first one is smaller than the second one except two positions of No. 5 and No. 6. The first peak of the separating frequency is shifted to close the resonant frequency (13.56 MHz) when the defected position moves away from the center. As the defected position moves outward to the center of the MTS, the |S_{21}| and the frequency of the first peak also slowly increased accordingly. At the defected position of No. 4, the magnitude of |S_{21}| gets the minimum of 0.106 at 13.56 MHz. From the results, thus, the defected MTS for the cases of the large transmitting and receiving coils has not improved the efficiency due to the magnetic field confinement on the defected cell, as shown in **Figure 5(c)**.

In practice, a transmitting side with a larger coil size than a receiving side for charging portable and implantable devices is used. The typical relative size ratio between transmitting and receiving coils is large, so the efficiency results low. The configurations for a large transmitting side and a small receiving side with uniform and defected MTSs are shown in **Figure 1(d)** and **Figure 2(b)**, respectively. A comparison of the |S_{21}| at six positions for uniform and defected MTSs is shown in **Figure 6**. It can be seen in **Figure 6(a)** for uniform MTS that the |S_{21}| has a significant deterioration when the receiving coil move outward from the center. The difference |S_{21}| at 13.56 MHz between position No. 1 (center) and No. 6 is about 0.42 or 10 dB down, when the maximum is 0.61 at the receiving position in the center. **Figure 6(c)** shows the |S_{21}| of both uniform and defected MTSs at each receiving position. For the uniform MTS, the magnitude gradually decreases from 0.61 to 0.19 when the receiving coil moves from the center (No. 1) to the edge (No. 6) of the MTS.

Meanwhile, the defected MTS shows a |S_{21}| from 0.64 to 0.57. This is confirmed that the proposed defected MTS provides a relatively constant transfer efficiency in all the areas of the MTS. In addition, the defected MTS provides increased transfer efficiency compared with the cases of the free space and uniform one. This proposed configuration is in contrast of the [20, 24], which is constructed in the

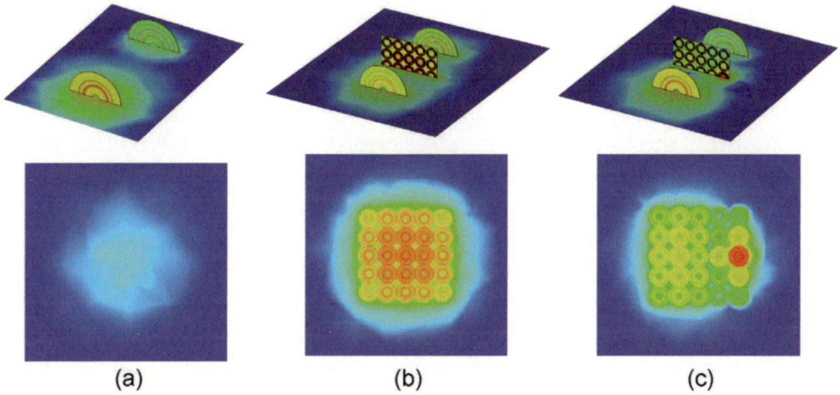

Figure 5.
Magnetic field intensity distribution for LTLR (a) free space, (b) uniform MTS when the receiver at the center and (c) defected MTS when the receiver located at No. 4.

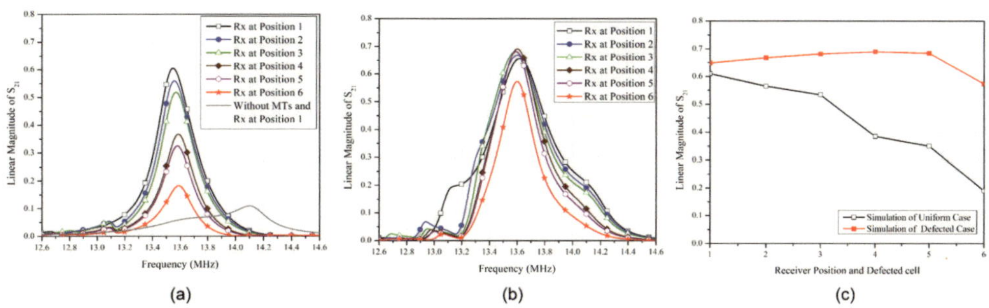

Figure 6.
Comparison of the $|S_{21}|$ for LTSR with different positions of the receiving coil on (a) the uniform MTS, (b) the defected MTS when the defected unit cell is the same position with the receiving coil and (c) both uniform and defected MTSs.

cavity using an array of the defected unit cells. So, the proposed defected MTS can enhance the transfer efficiency, even with its own loss.

The magnitudes of the magnetic field distribution are also used to compare between the WPT systems as shown in **Figure 5**. When whether the uniform or defected MTSs are inserted in the system, strong surface waves existing on both sides of the MTS are observed, which are responsible for the increased magnetic coupling. Obviously, the magnetic field intensity of free space is relatively weak in comparison with the metasurface cases. In case of free-space and the uniform MTS with the LTLR, the receiving coil is placed at the center. The intensity for both cases at the center is more concentrated than at the edge and the uniform MTS has a higher since the magnetic field is enhanced and focused by the negative permeability of the uniform MTS. In the case of the uniform MTS in **Figure 5(b)**, it is showed that the focusing effect of the magnetic field is present due to the negative refractive lens, which the field gradually decays from the center to the edge so, lead to better efficiencies. When the defected unit cell is selected at No. 4 as shown in **Figure 5(c)**, the focusing effect of the magnetic field is presented at position No. 4. However, For the other unit cell of the defected metasurface, the magnetic field intensity is significantly dropped compared with the case of the uniform MTS. According to the magnetic field distribution, the magnitude S_{21} value of the case of the defected metasurface is not too high, as the case of the uniform metasurface at a frequency of 13.56 MHz, as shown in **Figure 4** because the magnetic field intensity is confined only specific position.

Figure 7 shows the magnetic field intensity distributions of LTSR with the uniform and defected MTSs at 13.56 MHz, respectively. On the uniform MTS in **Figure 7(a)**, the magnetic field spreads over a relative area, while the defected MTS shows that the focusing effect of the magnetic field from the cavity is presented, which can suppress the filed elsewhere. It is observed in **Figure 7(b)** that the strong magnetic field confinement is realized on the defected MTS at the receiving position and the field is relatively low with uniform outside the cavity. It means that the defected MTS creates the field localization, hence, it can enhance transfer efficiency for a small receiver and reduce leakage EMF. It is because the unit cell forming the cavity resonance while the surrounding cells resonance at lower frequency. When the resonant frequency of surrounding cells falls into the hybridization bandgap, the negative permeability of metasurface forms a stopband for the cavity. Thus, the magnetic fields are prohibited from propagation in an outside area than the cavity. For the uniform MTS, the magnitude of the magnetic field is distributed relatively evenly about 14–22 dB along the surface of the MTS whereas the defected MTS is below 6 dB. Therefore, the defected MTS can enhance not only the field at the receiving coil, but also suppress the field elsewhere without additional shielding box or ferrite.

To experimentally validate the performance of the uniform and defected MTSs, several experimental studies are conducted. **Figure 8** shows the prototype of the proposed metasurface and experimental setup by using the VNA (Rohde & Schwarz model ZVB20). The fabricated MTS consists of 5 × 5 arrays of unit cells that are

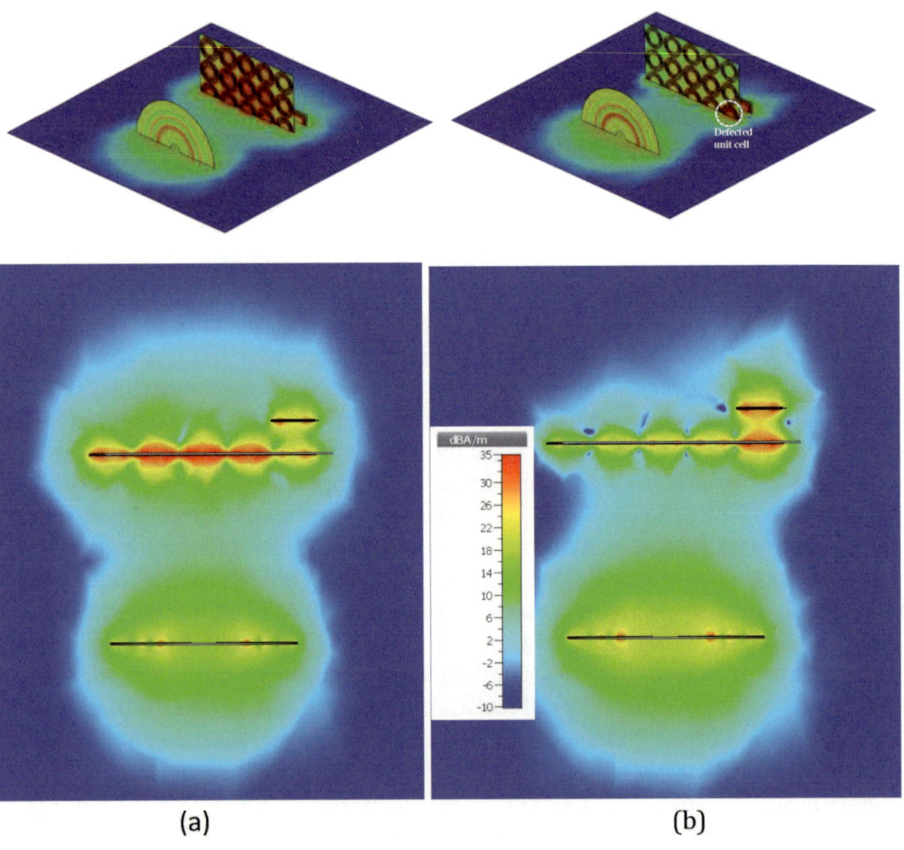

Figure 7.
Magnetic field intensity distribution for LTSR when the receiver located at No. 4 (a) uniform MTS and (b) defected MTS.

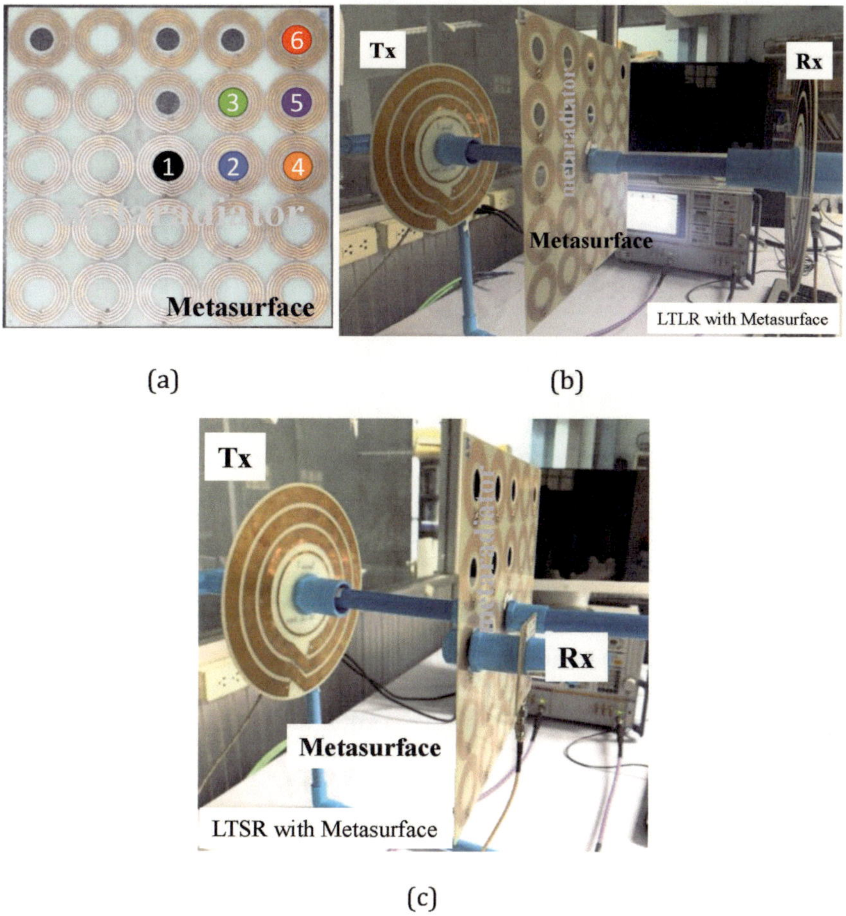

Figure 8.
Photograph of prototype of the proposed metasurface (a) and the system experimental setup for measurement: (b) LTLR and (c) LTSR.

controlled individually using a chip capacitor as shown in **Figure 8(a)**. **Figure 8(b)** depicts the LTLR with the metasurfaces. In case of LTLR (**Figure 8(b)**), the large receiving is not fixed at the center, but it is moved and located following the defected unit cell. **Figure 8(c)** shows the case of the LTSR with the metasurfaces when the receiving coil can move all the unit cell locations. The distance between the receiving coil and MTS is fixed and kept by using the plastic pipe. It is a thin and non-magnetic material; its effect is negligible. The connection is accomplished using standard SMAs through two identical low-loss cables. The standard SOLT (short-open-line-trough) calibration has been performed at the desired frequency range before measurement, so the end of the cables has been considered as a reference plane for S-parameters. **Figure 9** shows the measured results of LTLR and LTSR cases with the uniform and defected MTSs, respectively. **Figure 9(a)** shows variation of the measured efficiency depending on the defected positions and the defected cell at position No. 0 is a uniform MTS case. As seen, the efficiency shows a significant dependence on the receiving positions. The WPT system is directly measured at the operating frequency of 13.56 MHz and using two-port method with VNA. It is clearly seen in **Figure 9(b)** that the transfer efficiency of the defected MTS is quite flat; it means the defected MTS can enhance the efficiency regardless of the positions of the receiving coil with various distances. **Figure 9(b)** shows the simulated and measured power transfer efficiency of LTLR and LTSR cases with the uniform and defected

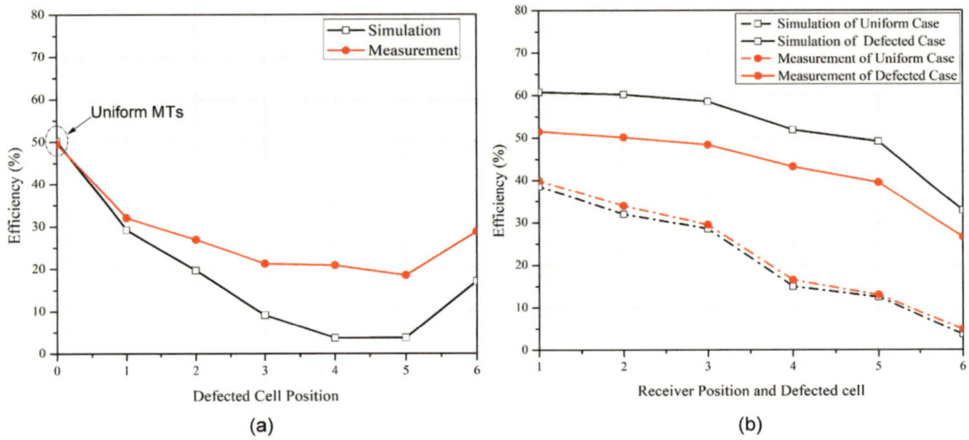

Figure 9.
Measured and simulated results of the transfer efficiency with the uniform or defected metasurface (a) LTLR and (b) LTSR.

MTSs. We obtained the measured transfer efficiency of LTSR case of 51,5%, 50.1%, 48.3%,43.2%, 39.5% and 26.6% at the receiver positions of No. 1 to No. 6, respectively. Contrastingly, for the uniform MTS, the efficiency decreases obviously, when the receiving positions are moved far away from the center since the intensity of magnetic field at the edge is lower than at the center for the uniform MTS.

4. Conclusions

In this chapter, we propose the defected MTS for enhancing transferred efficiency and reducing the EMF. The MTS is based on a defected unit cell at the desired receiving position formed two-dimensional cavity. It can improve the localization of the magnetic field; thus, the strong confinement of the receiving location can be realized. Compared to previous defected MTSs especially in case of the small transmitter, using this proposed method, it is a simple way to create the cavity of two-dimension area because the single unit cell is only used and controlled in contrast of the series of an array of the defected unit cells via hopping or MIW route. The power transfer efficiency of the proposed defected MTS increases significantly from 1.9% (free space case) and 38% (uniform MTS) to 60% with the size ratio between transmitter and receiver is 4:1. Moreover, the confinement can reduce the leakage EMF around the surrounding system without additional shielding box. It can reduce EMF leakage about 10 to 15 dB at 13.56 MHz with the defected MTS. When the defected MTS is integrated in the WPT system, however, the size of the receivers is affected of the overall efficiency. If the size of the receiver is comparable compared with the unit cell size, it gets better efficiency than the larger receivers. Consequently, the defected MTS is not suitable for the case of the large transmitting and receiving coils due to splitting frequency effect. We hope that thought the proposed defected MTS operates in low frequency, the proposed structure can be easily scaled and tuned by adjusting the geometry of the coil and changing the capacitor value.

Acknowledgements

This work was supported by the Thailand Research Fund under Grant RSA6280056.

Conflict of interest

The authors declare no conflict of interest.

Author details

Sarawuth Chaimool[1*], Chawalit Rakluea[2], Yan Zhao[3] and Prayoot Akkaraekthalin[2]

1 Electrical Engineering, Faculty of Engineering, Khon Kaen University, Khon Kaen, Thailand

2 Electrical and Computer Engineering, Faculty of Engineering, King Mongkut's University of Technology North Bangkok, Bangkok, Thailand

3 International School of Engineering, Faculty of Engineering, Chulalongkorn University, Bangkok, Thailand

*Address all correspondence to: sarachai@kku.ac.th; jaounarak@gmail.com

References

[1] iphone12 [Internet]. 2020. Available from https://support.apple.com/en-us/HT211829

[2] Qi standard [Internet]. 2020. Available from https://www.wirelesspowerconsortium.com/qi/

[3] Mi C, Buja G, Choi S, Rim C: Modern advances in wireless power transfer systems for roadway powered electric vehicles. IEEE Trans. Ind. Electron. 2016;10:6533-6545. DOI: 10.1109/TIE.2016.2574993.

[4] Patil D, McDonough M, Miller J, Fahimi B, Balsara P: Wireless power transfer for vehicular applications: overview and challenges. IEEE Trans. Transp. Elect. 2018; 4: 3-37. DOI: 10.1109/TTE.2017.2780627.

[5] Panchal C, Stegen S, Lu J:Review of static and dynamic wireless electric vehicle charging system: Int. J. Eng. Sci. Tech. 2018; 21:922-937. DOI: 10.1016/j.jestch.2018.06.015

[6] Garnica J, Chinga R, Lin J: Wireless power transmission: from far field to near field. IEEE Proc. 2013;101: 1321-1331. DOI: 10.1109/JPROC.2013.2251411.

[7] Corrêa D, Resende U, Bicalho F: Experiments with a compact wireless power transfer system using strongly coupled magnetic resonance and metamaterials. IEEE Trans. Magn. 2019;55: 8401904. DOI: 10.1109/TMAG.2019.2913767

[8] Huang R, Zhang B: Frequency, impedance characteristics and hf converters of two-coil and four-coil wireless power transfer. IEEE J. Emerg. Sel. Pow. Electron. 2015; 3:177-183. DOI:10.1109/JESTPE.2014.2315997.

[9] Cheon S, Kim Y, Kang S, Lee M, Lee J, Zyung T:Circuit-model-based analysis of a wireless energy-transfer system via coupled magnetic resonances. IEEE Trans. Ind. Electron. 2010;58: 2906-2914. DOI: 10.1109/TIE.2010.2072893.

[10] Hui S. Y. R, Zhong W, Lee C. K.: A critical review of recent progress in mid-range wireless power transfer. 2013;29: 4500-4511. DOI: 10.1109/TPEL.2013.2249670.

[11] ICNIRP. Guidelines for limiting exposure to electromagnetic fields (100 kHz to 300 GHz). Health Phys 118(00):000-000; 2020. Pre-print. DOI: 10.1097/HP.0000000000001210.

[12] Song C, Kim H, Jung D. H, Kim J, Kong S, Kim J, Ahn S, Kim J, Kim J: Low EMF and EMI design of a tightly coupled handheld resonant magnetic field (HH-RMF) charger for automotive battery charging. IEEE Trans. Electromag. Compat. 2016; 58: 1194-1206. DOI: 10.1109/TEMC.2016.2557842.

[13] Park H, Kwon J, Kwak S, Ahn S: Effect of airgap between a ferrite plate and metal strips on magnetic shielding. IEEE Trans. Magn. 2015; 51: 9401504. DOI: 10.1109/TMAG.2015.2432102.

[14] Li J, Yin F, Wang L, Wang L: Research on the transmission efficiency of different shielding structures of wireless power transfer system for electric vehicles. CSEE J. Pow. Ener. Sys. 2020; DOI: 10.17775/CSEEJPES.2019.00500.

[15] Ranaweera A, Moscoso C, Lee J: Anisotropic metamaterial for efficiency enhancement of mid-range wireless power transfer under coil misalignment. J. Phys. D: Appl. Phys. 2015; 48:455104(8pp). DOI:10.1088/0022-3727/48/45/455104.

[16] Younesiraad H, Bemani M: Analysis of coupling between magnetic dipoles

enhanced by metasurfaces for wireless power transfer efficiency improvement. Sci. Reports. 2018; 8:14865. DOI: 10.1038/s41598-018-33174-8.

[17] LI L, Liu H, Zhang H, Xue W: Efficient wireless power transfer system integrating with metasurface for biological applications. IEEE Trans. Indus. Electron. 2018; 4:3230-3239. DOI: 10.1109/TIE.2017.2756580.

[18] Rakluea C, Chaimool S, Zhao Y, Akkaraekthalin P: Compact non-uniform metasurface for efficiency enhancement of planar wireless power transfer. In: Proceedings of the 2019 International Electrical Engineering Congress (iEECON2019), 6-8 March 2019, Cha-am, Thailand. New York: IEEE; 2019. p.1-4

[19] Lee W, Yoon Y: Wireless power transfer systems using metamaterials: A review. IEEE Access. 2020; 8: 147930-147947. DOI: 10.1109/ ACCESS.2020.3015176.

[20] Bui H, Pham T, Ngo V, Lee J: Investigate of various cavity configurations for metamaterial-enhanced field-localizing wireless power transfer. J. Appl. Phys. 2017; 122:093102-093110. DOI: 10.1063/1.5001130.

[21] Lu C, Huang X, Rong C, Hu Z, Chen J, Tao X, Wang S, Wei B, Liu M: Shielding the magnetic field of wireless power transfer system using zero-permeability metamaterial. J. Eng. 2019; 16:1812-1815. DOI: 10.1049/ joe.2018.8678.

[22] Cho Y, Lee S, Kim D, Kim H, Song C, Kong S, Park J, Seo C, Kim J: Thin hybrid metamaterial slab with negative and zero permeability for high efficiency and low electromagnetic field in wireless power transfer systems. IEEE Trans. Electromagn. Compat. 2017; DOI: 10.1109/ TEMC.2017.2751595.

[23] Bui H, Kim J, Lee J: Design of tunable metasurface using deep neural networks for field localized wireless power transfer. IEEE Access. 2020; 8:194868-194878. DOI: 10.1109/ Access.2020.3033527.

[24] Bui H, Pham T, Kim J, Lee J: Field-focused reconfigurable magnetic metamaterial for wireless power transfer and propulsion of an untethered microrobot. J. Magn. Mater. 2020; 494:165778. DOI: 10.1016/j. jmmm.2019.165778.

[25] Pham T, Ranaweera A, Lam V, Lee J: Experiments on localized wireless power transmission using a magneto-inductive wave two-dimensional metamaterial cavity. Appl. Phys. Exp. 2016; 9:044101. DOI: 10.7567/APEX.9.044101.

[26] Shamonica E, Kalinin V, Solymar L: Magnetoinductive waves in one, two, and three dimensions. J. Appl. Phys. 2002; 92:6252-6261. DOI:10.1063/1.1510945.

[27] Sandoval F, Delgado S, Moazenzadeh A, Wallrabe U: A 2-D magnetoinductive wave device for freer wireless power transfer. IEEE Trans. Power Electron. 2019; 34: 10433-10445. DOI: 10.1109/TPEL.2019.2904875.

[28] Stevens C: Magnetoinductive waves and wireless power transfer. IEEE Trans. Power Electron. 2014; 30: 6182-6190. DOI: 10.1109/TPEL.2014.2369811.

[29] Campione S, Mesa F, Capolino F: Magnetoinductive waves and complex modes in two-dimensional periodic arrays of split ring resonators. IEEE Trans. Antennas Propag. 2013; 61: 3554-3563. DOI: 10.1109/ TAP.2013.2258395.

[30] Pham T, Ranaweera A, Ngo D, Lee J:Analysis and experiments on Fano interference using a 2D metamaterial cavity for field localized wireless power transfer. J. Phys. D: Appl.

Phys. 2017; 50:305102(10pp). Doi: 10.1088/1361-6463/aa7988.

[31] Chaimool S, Rakluea C, Akkaraekthalin P, Zhao Y: Effect of losses in printed rectangular coils for compact wireless power transfer systems. Prog. Electromag. Research C. 2019; 97:177-188. DOI:10.2528/PIERC19092601.

[32] CST Microwave Studio, Computer Simulation Technology, Framing-ham, MA, 2015.

[33] Ranaweera A, Pham T, Bui N, Ngo V, Lee J: An active metasurface for field-localizing wireless power transfer using dynamically reconfigurable cavities. Sci. Rep. 2019; 9:11735. DOI: 10.1038/s41598-019-48253-7.

[34] Lyu Y, Meng F, Yang G, Che B, Wu Q, Sun L, Ermi D, Li J:A method of using nonidentical resonant coils for frequency splitting elimination in wireless power transfer. IEEE Trans. Power Electron. 2015; 30:6097-6107. DOI: 10.1109/TPEL.2014.2387835.

[35] Lv B, Li R, Fu J, Wu Q, Zhang K, Chen W, Wang Z, Ma R: Analysis and modeling of Fano resonances using equivalent circuit elements. Sci. Rep. 2016; 6:31884. DOI: 10.1038/srep31884.